T0295744

Water Management in China's Power Sector

This book examines water resource management in China's electric power sector and the implications for energy provision in the face of an emerging national water crisis and global climate change.

Over 75% of China's current electricity comes from coal. Coal-fired power plants are reliant on water, with plants using significant volumes of water every year, yet water resources are unevenly distributed. In the face of serious environmental concerns and increasing electricity demand, this book examines the environmental impacts that coal power plants have on water resources and the impact water availability has on the electricity sector in a country with a significant number of water-scarce provinces and a large number of power plants located on inland waterways. It discusses the water impacts and constraints for transforming the electric power sector away from coal to renewable energy sources, such as hydropower and concentrated solar power. The book adopts a mix-method approach, combining a plant-level quantitative analysis on water impacts and dependencies in China's electricity sector and a qualitative analysis of relevant institutions in both sectors. By reviewing policy and institution cases in China's water and electricity sectors, the book provides important recommendations calling for coordinated institutions to shift away from the current paradigm where water and electricity are governed independently.

Enriching the water-energy nexus literature, this book will be of great interest to students and scholars working on water resource management, energy industries and Chinese environmental policy, as well as policymakers and practitioners in those fields.

Xiawei Liao is a water resource management specialist and researcher. He is currently a consultant for the World Bank Water Global Practice and a postdoctoral researcher at Peking University, China.

Jim W. Hall is Professor of Climate and Environmental Risks in the University of Oxford, UK, where he is Director of Research in the School of Geography and the Environment.

Earthscan Studies in Water Resource Management

Water, Technology and the Nation-State
Edited by Filippo Menga and Erik Swyngedouw

Revitalizing Urban Waterway Communities
Streams of Environmental Justice
Richard Smardon, Sharon Moran and April Baptiste

Water, Creativity and Meaning
Multidisciplinary Understandings of Human-Water Relationships
Edited by Liz Roberts and Katherine Phillips

Water, Climate Change and the Boomerang Effect
Unintentional Consequences for Resource Insecurity
Edited by Larry Swatuk and Lars Wirkus

Legal Rights for Rivers
Competition, Collaboration and Water Governance
Erin O'Donnell

Water Allocation Law in New Zealand
Lessons from Australia
Jagdeepkaur Singh-Ladhar

Reciprocity and China's Transboundary Waters
The Law of International Watercourses
David J. Devlaeminck

Water Management in China's Power Sector
Xiawei Liao and Jim W. Hall

For more information about this series, please visit: www.routledge.com/books/series/ECWRM/

Water Management in China's Power Sector

Xiawei Liao and Jim W. Hall

First published 2021
by Routledge
2 Park Square, Milton Park, Abingdon, Oxon OX14 4RN

and by Routledge
52 Vanderbilt Avenue, New York, NY 10017

Routledge is an imprint of the Taylor & Francis Group, an informa business

British Library Cataloguing-in-Publication Data
A catalogue record for this book is available from the British Library

Library of Congress Cataloging-in-Publication Data
A catalog record for this book has been requested

ISBN: 978-0-367-35005-5 (hbk)
ISBN: 978-0-429-34659-0 (ebk)

Typeset in Times New Roman
by Apex CoVantage, LLC

Contents

Figures

Tables

1 Introduction

Electricity (hereinafter used interchangeably with 'electric power' and 'power') provision underpins modern human society's development and prosperity. From the UNDP Human Development Index and World Bank electricity consumption per capita data (Figure 1.1), it can be seen that, especially in the early stage of development, electricity use is highly correlated with the development of human society, which is not difficult to comprehend as electricity provides a wide range of fundamental services, from domestic uses, such as lighting and heating, to industrial production, commerce and government services as well as others (e.g. vaccine storage in the health sector and electric vehicles in the transport sector) that enable the society to operate, develop and prosper.

China has made remarkable progress in its social-economic development. It has lifted over 850 million of its population out of poverty during the last 40 years since its 'Reform and Opening Up' policy in 1978. Its economy has maintained rapid growth at an average annual growth rate of over 10% from 1978 to 2007 and slowed down during the last ten years, entering its so-called economic new normal. It has become the second-largest economy in the world, with a share in the world economy having increased from 1.5% in 1978 to over 15% in 2017. The income per capita has increased fortyfold from 200 USD to 8,690 USD in the same period. This economic transformation has been fueled by soaring electricity consumption. Entering the 21st century, China's national electricity consumption has more than quadrupled from 1347 TWh in 2000 to 5802 TWh in 2015 and has overtaken the US as the biggest electricity consumer, as well as producer, in the world (National Bureau of Statistics of China, 2016). It should be noted that, dividing by population, China's per capita electricity consumption is still very low at only 4231 KWh per person in 2015, compared with an average of 7,995 KWh per person in OECD countries and 12,987 KWh per person in the US (World Bank, 2017).

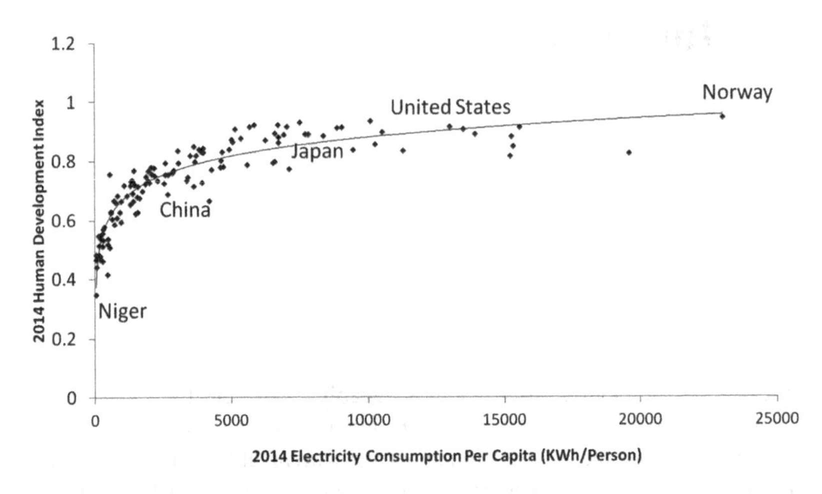

Figure 1.1 Global Human Development Index and electricity consumption per capita (kWh) in 2014

Source: UNDP, World Bank.

With the third-largest coal reserves in the world, China is the largest producer and consumer of coal in the world and is the largest user of coal-derived electricity (World Bank, 2017). As of the end of 2016, China had over 978 GW of coal-fired power plant capacity, which made up 57% of the national total electricity capacity. China has another 227 GW of coal-fired power plants under construction and 563 GW in various planning stages. In terms of the electric power output, it can be seen from Figure 1.2, although with a declining proportion, over 70% of China's electricity was still generated by thermoelectric power plants in 2015, among which coal power plants account for more than 90%. Hydropower provides 15–20% of the electricity for China, primarily in water-abundant southwestern China. In recent years, there has been a rapid growth of renewable electricity, primarily from solar photovoltaic (PV) panels and wind, though this still only represented around 5% of power production in 2015. The proportion of nuclear power production has also increased from under 2% in 2010 to just below 4% in 2016.

The utilization of coal is facing increasing challenges. The harmful impacts of air pollution and greenhouse gas emissions from coal power generation has gained increasing publicity and attention. Burning coal has been recognized as a major source of air pollutants, particularly sulfur dioxide, nitrogen oxides, particulate matters and so on. According to estimates from the Global Burden of Disease study (2017), air pollution is the fourth

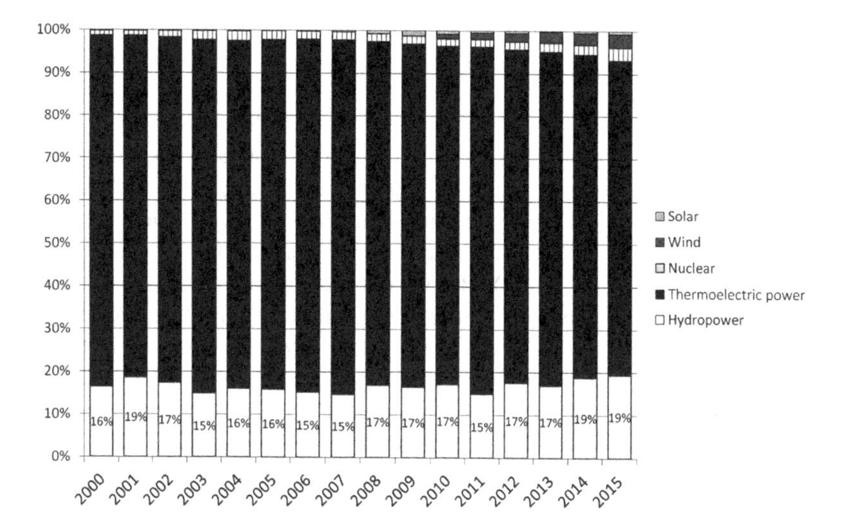

Figure 1.2 Energy structure of China's electric power sector

leading cause of death and disability in China. In recent years, hazy air, caused by particular matters PM2.5 and PM10, has given rise to increasing social unrest in China. It has been recognized that the scale and complexity of China's environmental problems require a fundamentally new approach to a growth model that allows economic growth and development while decoupling upward trends in resource use and environmental degradation. Transitioning the Chinese economy from polluting technologies to more efficient, low-carbon and cleaner development trajectory would require a new set of policies, regulatory and institutional frameworks.

The 13th Five-Year Plan (2016–2020) of China and the 19th Chinese Communist Party Congress Report (October 2017) called for a 'beautiful China' founded by pursuing productivity and innovation-driven development; continuing to rebalance toward consumption and services and reversing environmental degradation. In response to record-high levels of air pollution in 2012 and 2013, the State Council of China issued an 'Action Plan for the Prevention and Control of Air Pollution' in September 2013 and reiterated the need to reduce coal's share in China's primary energy mix to 65% by 2017. Meanwhile, increasingly stringent environmental regulations have also significantly contributed to the reduction of other air pollutants. For example, since the 'Flue Gas Limestone/Limegypsum desulfurization project technical specification of thermal power plant' was issued in 2005, desulfurization facilities have become widely applied in China's coal power plants (almost 100% of China's coal power plants are equipped with desulfurization facilities now), sulfur dioxide emission intensity has been reduced from 6.4 g/KWh to 2.26 g/KWh, even lower than that in the US (Li and Chen, 2013).

On the other hand, since China overtook the US as the world's largest carbon emitter in 2006, its coal power sector, a main source for greenhouse gas emissions, has come under increasing scrutiny. In 2014, China's carbon emissions made up about 28.8% of the world total, at 10.4 billion tons, among which coal power generation has contributed more than 30% (Liu, 2015). Before the 2009 United Nations Climate Change Conference in Copenhagen, China made a commitment to reduce its carbon emissions intensity per GDP by 40 to 45% until 2020. However, the lack of commitment on total carbon emissions control has drawn much criticism. In China's 13th Five-Year Plan (2016–2020) (National Development and Reform Commission, 2016), it further pledged to peak its carbon emissions by 2030 and reduce the share of coal to below 60% in its primary energy structure by 2020 and further cut it down to 50% by 2030.

While China's increasing demand and production of coal-derived electricity has triggered growing concern regarding its environmental externalities, as mentioned previously, those concerns mostly have been focused on air pollutions and greenhouse gas emissions from burning fossil fuels. The use of and impacts on another essential natural resource (i.e. water) by coal power production has received less attention (Gleick, 1994). Water is used throughout the lifecycle supply chain to provide coal-derived electricity, from upstream coal mining and washing to power plant on-site usage, primarily for cooling uses. It is estimated that thermoelectric cooling is responsible for around 43% and 50% of total freshwater withdrawals in the EU and US, respectively (UN, 2014).

Water is a fundamental resource that often goes underappreciated. People start to pay close attention only when there is either too much or too little of it (Sadoff et al., 2015). Driven by a growing population and economic activity, the risks to water resources to meet human needs, while maintaining the aquatic environment, has been identified as a top global risk facing humanity over the next decade (World Economic Forum, 2015). As coal power plants worldwide depend on large amounts of water for cooling every year, both water shortage and water temperature increase could generate negative impacts on electricity generation, which is manifested by the curtailments of power production or even plant shutdowns in Europe, for instance, as seen in 2003 in France. From 2000 to 2015, 43 thermoelectric power curtailments because of water-related issues have been documented in the Unites States (McCall, Macknick and Hillman, 2016). In addition, under a changing climate, the availability and variability of water resources are also expected to change, both geographically and seasonally, which may further intensify the aforementioned risks (Van Vliet et al., 2012, 2016; Zheng et al., 2016).

At present, almost one-fifth of the global population, around 1.2 billion people, are living in conditions of physical water scarcity, where water resources are insufficient to support human society. In addition, another 1.6 billion people are facing economic water shortages, where there is enough water in the natural system but a lack of sufficient infrastructure to withdraw the water for humans to use (UN and FAO, 2007). Furthermore, it is expected that, by 2030, water demand will exceed supply by 40% (UN WWAP, 2019), and by 2050, water demand will further increase by 55% (OECD, 2012). Under a baseline scenario, the number of people living with severe water stress is projected to reach 3.9 billion, more than double the current level (OECD, 2012). Therefore, if the business continues as usual, especially in some water-scarce regions, electricity production is set to compete with other basic human water needs (e.g. drinking, food, sanitation) and the ecological system for limited water resources in the hydrologic system (Bazilian et al., 2011; Byers et al., 2016).

The aforementioned impacts and risks are particularly pronounced in China because of the prevalence of thermoelectric power plants on inland waterways that are potentially at risk of water shortages, which is further exacerbated by the uneven geographical distribution of China's water endowments and their mismatch with energy reserves, particularly coal (Zhang and Anadon, 2013). First of all, China is not abundant in water. As it is home to nearly 20% of the global population, China is endowed with only 6% of the world's freshwater resources, leaving its annual renewable freshwater per capita level at about only one-third of the global average, at around 2100 m^3 per person (Ye and Zhang, 2013). Furthermore, the geographical distribution of China's water resources is highly uneven. North China is home to 37% of its national population and 45% of the arable land but only 12% of the water resources. Eleven provinces out of 31 in mainland China are facing water stresses, mainly in the north. More than 650 million people now live in water-stressed areas in China, and the number is on the rise. What makes it worse is that China's water resources mismatch its energy reserves. The challenges of this energy-water conflict are highlighted by the location of an estimated 71% of China's coal resources in four water-scarce arid and semi-arid regions of northern China that account for only 13% of total water resources. Water-scarce provinces (with water resources similar to the Middle East) generate nearly half of China's GDP and hold over half of China's ensured coal reserves. To save coal transportation costs, a large proportion of the power generation capacity is located in these areas close to the coal reserves, as well as over half of the country's proposed coal-fired power plants (China Water Risk, 2012; World Resource Institute, 2014). According to the assessment by Sadoff et al. (2015), northern China stands out as having the highest water security risks in the world for its power generation. Limited water availability may constrain the development of the electricity sector, or at least require more costly technologies.

Water constraints have already impeded energy developments in China. In response, in 2004, the National Development and Reform Commission issued 'Requirements on the Planning and Construction of Coal Power Plants' and required new coal-fired power plants in the water-scarce regions to use air-cooling systems and encouraged plants in other regions to do the same. Groundwater utilization in water-scarce regions was prohibited by the same document. In 2008, plans to build dozens of coal-to-liquid (CTL) plants were abandoned because of local water scarcity (IEA, 2012).

These policies reflect a larger transition of China aiming to promote greener growth. After a period of unprecedented economic growth and social development, China is embarking upon a transition to a more balanced and sustainable economic growth model. This transition is centered in part on sustainable resource management, environmental protection and ecological conservation, with sustainable water management central to the realization of China's sustainable economic prosperity. In 2011, the Chinese government issued its strictest water management policy, called the 'Three Red Lines' policy (State Council, 2012). This policy set targets for total water use, industrial and agricultural water use efficiencies and water quality improvements on a national as well as on a regional scale for 2015, 2020 and 2030. Regarding the total water use, the first 'Red Line' sets a national water withdrawal cap at 635, 670 and 700 billion m^3 for 2015, 2020 and 2030, respectively, against the level of 609.5 billion m^3 in 2014 (China Water Resource Bulletin, 2015), which will be enforced through a water permit system. Furthermore, recognizing the water impacts of the coal sector, especially in China's dry north, two years later, the Chinese government issued an addition to the 'Three Red Lines' policy specifically for the coal sector (i.e. 'Water Allocation Plan for the Development of Coal Bases'). The plan requires the coal sector to reduce its water usage, improve water efficiency and reduce wastewater discharges. Future large-scale coal projects, especially the ones in water-scarce regions, are required to be developed in coordination with local water authorities (Qin et al., 2015).

Another way to alleviate local water pressure imposed by electricity production is to import electricity from other areas. Whereas water is highly place-specific but unevenly distributed in China, electricity is increasingly being transmitted. As Bai (2016) has pointed out, any anthropogenic system depends on natural resources not only locally but also externally. Water can be redistributed in the form of virtual water embodied in goods transmitted across geographical boundaries (Zhao et al., 2015). These interdependencies mean electricity consumption in one region not only has impacts on its local water resources but also has impacts beyond its territories (Liu et al., 2015). Consequentially, electricity consumption also faces water-related risks occurring in other regions.

However, it is evident that water management issues are not always on the agenda of energy policymakers. For example, China's coal-fired power plants have shown an increasing trend of over-capacity. In 2016, the utilization rates of coal power facilities reached the lowest point since 1969, 49.4%. Against the backdrop of coal power plants' over-capacity, in 2016, the Chinese government took various measures to cut down hundreds of GW of planned or proposed coal power plants; even the ones that are already under construction are postponed (Caldecott et al., 2016). Following these measures, the National Energy Administration of China issued a 'Notice on Establishing a Coal-fired Power Planning and Construction Risk Warning Mechanism (hereinafter referred to as 'the mechanism')' in 2016 (National Energy Administration, 2016). The mechanism grades each province for its suitability to further expand its coal-fired power capacity from Red (discouraged development), Orange (cautious development) to Green (normal development) based on three sets of indexes: bankability, generation capacity adequacy and resource constraints. Although the mechanism explicitly listed water availability as one of the resource constraints, development of coal-fired power plants was nonetheless not discouraged in several water-scarce provinces (e.g. Ningxia, Shanxi, Inner Mongolia).

Within the contexts of China's sustainability transition and increasing water stress, this book aims to provide a full picture of water management issues in China's power sector and to demonstrate how China's energy transition and increasing water stress can affect each other. Chapter 2 describes in detail how water is used in each process within a coal-fired power plant and what the major influencing factors are of the water intensity (m^3/MWh), water use per unit of electricity produced. Chapter 3 introduces and summarizes water use in the upstream stages of the coal power sector (e.g. coal mining and washing); Chapter 4 describes water uses in China on a high-resolution plant level; Chapter 5 demonstrates situations where available water resources in the river system are insufficient to meet the water uses described in Chapter 4. Contingencies brought by future climate change will also be discussed; Chapter 6 investigates the impact of inter-provincial electricity transmission on water resources, as well as the redistribution of water resources in the form of virtual water embodied in electricity transmitted; Chapter 7 reviews water uses of other alternative energy sources to coal for electricity production (i.e. hydropower, concentrated solar power and nuclear power) and the potential water constraints they may face if they are to be promoted to phase out coal gradually; Chapter 8 gives an in-depth and empirical analysis of how policies in both sectors in China could generate spill-over or unintended impacts on the other; Chapter 9 analyzes and presents the opportunities for and obstacles to institutional change that enables energy development that is coordinated with the water sector from the planning stage; Finally, Chapter 10 provides a conclusion.

Overall, this book aims to enrich the water-energy nexus literature by developing an in-depth perspective on one remarkable country. Insights in this book offer global implications for other developing countries in order to develop sustainable and water-sensitive energy systems to reduce impacts on water resources and to build resilience to water-related risks, especially under a changing climate.

References

Bai, X. (2016). Eight energy and material flow characteristics of urban ecosystems. *Ambio*, 45, pp. 819–830.

Bazilian, M., Rogner, H., Howells, M., Hermann, S., Arent, D., Gielen, D., Steduto, P., Mueller, A., Komor, P., Tol, R. S. J. and Yumkella, K. K. (2011). Considering the energy, water and food nexus: Towards an integrated modeling approach. *Energy Policy*, 39(12), pp. 7896–7906.

Byers, E. A., Hall, J. W., Amezaga, J. M., O'Donnell, G. M. and Leathard, A. (2016). Water and climate risks to power generation with carbon capture and storage. *Environmental Research Letters*, 11, p. 024011.

Caldecott, B., Dericks, G., Tulloch, D. J., Liao, X. W., Bouveret, G., Kruitwagen, L. and Mitchell, J. (2016). *Stranded assets and thermal coal in China: An analysis of environmental-related risk exposure*. Oxford: SSEE.

China Water Resource Bulletin. (2015). *The ministry of water resources of the People's Republic of China* (in Chinese). Beijing, China: China Water Resource Bulletin.

China Water Risk. (2012). *China: No water, no power*. Available at: http://chinawaterrisk.org/resources/analysis-reviews/china-no-water-no-power/.

Gleick, P. H. (1994). Water and energy. *Annual Review of Energy and the Environment*, 19, pp. 267–299.

The Institute for Health Metrics and Evaluation. (2017). *Global burden of disease study*. Available at: www.healthdata.org/china.

International Energy Agency (IEA). (2012). *World energy outlook*. Paris, France: IEA.

Li, X. and Chen, J. (2013). Pollution control of coal-fired power generation in China: An interview with Wang Zhixuan. *Cornerstone*. Available at: http://cornerstonemag.net/pollution-control-of-coal-fired-power-generation-in-china-an-interview-with-wang-zhixuan/.

Liu, J., Mooney, H., Hull, V., Davis, S. J., Gaskell, J., Hertel, T., Lubchenco, J., Seto, K. C., Gleick, P., Kremen, C. and Li, S. (2015). Systems integration for global sustainability. *Science*, 347(6225). doi:10.1126/science.1258832.

Liu, Z. (2015). *China's carbon emissions report*. Cambridge: Harvard University Press.

McCall, J., Macknick, J. and Hillman, D. (2016). *Water-related power plants curtailments: An overview of incidents and contributing factors*. Golden, CO: National Renewable Energy Laboratory.

National Bureau of Statistics of China. (2016). *Chinese energy statistics yearbook* (in Chinese). Beijing, China: National Bureau of Statistics of China.

National Development and Reform Commission of China. (2004). *Requirements on the planning and construction of coal power plants* (in Chinese). Available at: www.nea.gov.cn/2012-01/04/c_131262602.html.

National Development and Reform Commission of China. (2016). *13th five-year planning of the electric power sector (2016–2020)*. Beijing, China: National Development and Reform Commission of China.

National Energy Administration. (2016). *Notice on establishing a coal-fired power planning and construction risk warning mechanism*. Beijing, China: National Energy Administration.

Organisation for Economic Co-Operation and Development (OECD). (2012). *OECD environmental outlook to 2050: The consequences of inaction*. Paris, France: OECD Publishing.

Qin,Y., Curmi, E., Kopec, G. M., Allwood, J. M. and Richards, K. S. (2015). China's energy-water nexus assessment of the energy sector's compliance with the "3 red lines" industrial water policy. *Energy Policy*, 82, pp. 131–143.

Sadoff, C. W., Hall, J. W., Grey, D., Aerts, J. C. J. H., Ait-Kadi, M., Brown, C., Cox, A., Dadson, S., Garrick, D., Kelman, J., McCornick, P., Ringler, C., Rosegrant, M., Whittington, D. and Wiberg, D. (2015). *Securing water, sustaining growth: Report of the GWP/OECD task force on water security and sustainable growth*. Oxford: University of Oxford Press.

State Council of China. (2012). *Opinions on implementing the strictest water resources management system*. Beijing, China: State Council of China.

State Council of China. (2013). *Action plan for the prevention and control of air pollution*. Beijing, China: State Council of China.

UN and FAO. (2007). *Coping with water scarcity: Challenge of the twenty-first century*. Beijing, China: UN and FAO.

United Nations World Water Assessment Programme (UN WWAP). (2014). *The United Nations world water development report 2014: Water and energy*. Paris, France: UN WWAP.

United Nations World Water Assessment Programme (UN WWAP). (2019). *The United Nations world water development report 2019*. Paris, France: UN WWAP.

Van Vliet, M. T. H., Wiberg, D., Leduc, S. and Riahi, K. (2016). Power-generation system vulnerability and adaptation to changes in climate and water resources. *Nature Climate Change*, 6, pp. 375–380.

Van Vliet, M. T. H., Yearsley, J. R., Ludwig, F., Vögele, S., Lettenmaier, D. P. and Kabat, P. (2012). Vulnerability of US and European electricity supply to climate change. *Nature Climate Change*, 2, pp. 676–681.

World Bank. (2017). *World Bank databank*. Washington, DC. Available at: https://data.worldbank.org/indicator/EG.USE.ELEC.KH.PC [Accessed Nov. 2017].

World Economic Forum. (2015). *Global risks 2015*. Geneva, Switzerland: World Economic Forum.

World Resource Institute. (2014). *Identifying the global coal industry's water risks*. Available at: www.wri.org/blog/2014/04/identifying-global-coal-industrypercent E2percent80percent99s-water-risks.

Ye, W. H. and Zhang, Y. (2013). *Environment management*. 3rd Version (in Chinese). Beijing, China: China Higher Education Press.

Zhang, C. and Anadon, L. D. (2013). Life cycle water use of energy production and its environmental impacts in China. *Environmental Science & Technology*, 47(24), pp. 14459–14467.

Zhao, X., Liu, J., Liu, Q., Tillotson, M. R., Guan, D. and Hubacek, K. (2015). Physical and virtual water transfers for regional water stress alleviation in China. *Proceedings of the National Academy of Sciences of the United States of America*, 112, pp. 1031–1035.

Zheng, X., Wang, C., Cai, W., Kummu, M. and Varis, O. (2016). The vulnerability of thermoelectric power generation to water scarcity in China: Current status and future scenarios for power planning and climate change. *Applied Energy*, 171, pp. 444–455.

2 Engineering background

Overview of operational water use at coal power plants

As shown by the schematic of a coal-fired power plant in Figure 2.1, high-temperature and high-pressure steam is produced by burning coal in a furnace and converting the chemical energy stored in the coal into thermal energy that produces steam through the boiler. The high-temperature and high-pressure steam then drives the turbine and converts the energy into mechanical shaft motive energy, which drives the electric generator and eventually produces electricity. After exiting the turbine, the steam is routed to a condenser, and the condensed water is then pumped back into the boiler, repeating the cycle. The condensation process usually uses a cooling medium (i.e. water and air) to dissipate the residual heat carried in the exiting steam.

As can be seen from Figure 2.1, various processes at a coal-fired power plant require water as an input, for example, to clean and then transfer the coal, to remove the lime-ash and to desulfurize the flue gas, among which the largest amount of water is used to cool down the steam exiting the turbines. The water-using processes in a typical coal-fired power plant can be categorized as follows:

1 Industrial water use: Raw water withdrawn from the natural environment (e.g. river and ocean) first undergoes clarification and then can be used for different industrial purposes, such as ash removal and wet desulfurization, as well as cooling of the steam that drives the turbine and other auxiliary engines (including induced air fan, forced draft fan, primary air fan and so forth). Flue gas desulfurization systems use water combined with limestone or other agents to treat the flue gas to remove sulfur and lower SO_2 emissions. Furthermore, coal transportation also often uses water to create slurry that is a mixture of crushed coal suspended in liquid, usually water, as a means of transporting coal.

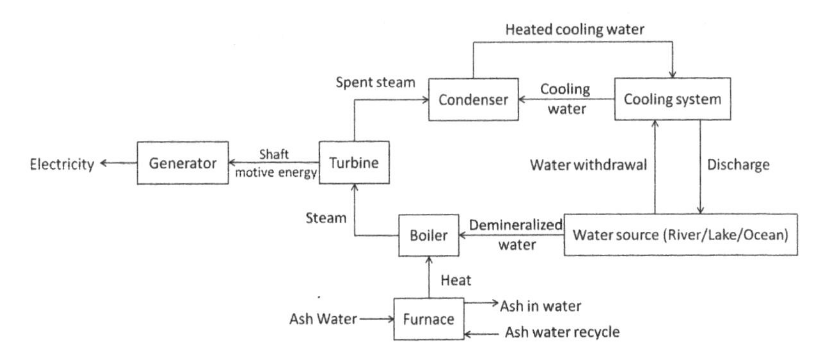

Figure 2.1 Water and energy flows in a coal-fired power plant

Source: Stone et al., 1982.

2 Domestic water use: Water is required to meet the living needs of power plant workers for drinking, cooking and sanitation services.
3 Boiler water use: Boiler needs to be regularly blown down to reduce the accumulating density of salt and other dissolved and suspended solids, which may cause foaming, priming and corrosion in the boiler system if not properly removed.

Two concepts, water withdrawal and water consumption, should be differentiated before proceeding further. The quantity of water withdrawn from the natural environment (e.g. rivers, lakes and oceans) is known as 'water withdrawal'. That water may be discharged back into the natural environment. Water that is withdrawn but not discharged back (because, for example, it is emitted into the atmosphere as steam) is defined as 'water consumption' (AQUASTAT, 1998). The term 'water use' is used generally to refer to water consumption or water withdrawal. Water intensity is defined as water use (m^3) per unit of electricity produced (KWh).

To understand cooling water uses, the three types of cooling systems currently being used should first be explained: open-loop and closed-loop wet cooling, and air-cooling:

Open-loop, or 'once-through' cooling (Figure 2.2)**:** If the power plant is located close to a large body of water, either to the sea or a big river, open-loop cooling systems can be used to run a large amount of cooling water through the condenser in a single pass and discharge the water withdrawn back into the environment a few degrees warmer. A very small amount of water will be consumed. Open-loop cooling systems withdraw the largest amount of water, but more than 99% is returned to the water source.

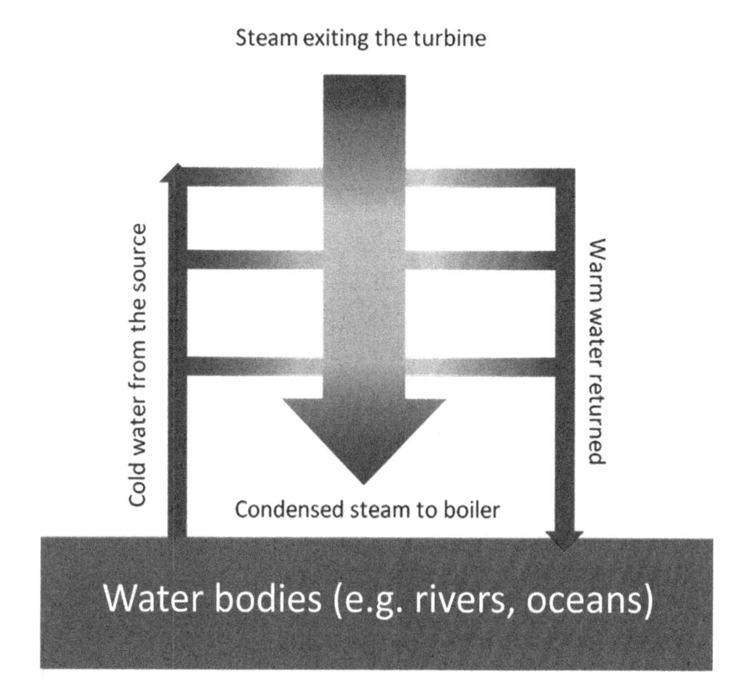

Figure 2.2 Water flows in open-loop cooling

Closed-loop, or recirculating cooling and cooling tower (Figure 2.3)

If the coal power plant is not built in water-abundant areas, a closed-loop cooling system can be employed that recirculates the cooling water in a closed loop where upward flowing air through water droplets cools the water. The cooling water is then condensed and can be reused. A large proportion of water withdrawn is eventually lost due to evaporation. As shown in Table 2.1, consumptive water use in a power plant with an open-loop cooling system is 70 to 80% lower than one with a closed-loop cooling system. However, open-loop cooling systems withdraw 30 to 60 times more water than closed-loop cooling systems (World Resource Institute, 2015).

Air-cooling systems, also called dry cooling (Figure 2.4)

In very arid places, air-cooling systems can be used to cool steam exiting turbines in a coal power plant. Without relying on evaporation, air-cooling

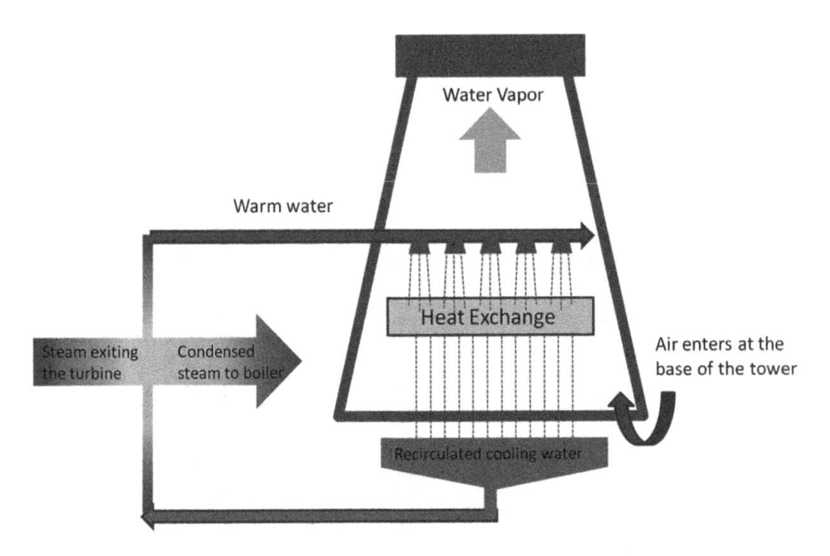

Figure 2.3 Water flows in closed-loop cooling

systems use cooling towers with a closed circuit, or high-forced draft air flow through a finned assembly. Power plants with air-cooling systems have the lowest water consumption and withdrawal intensities as they lose negligible quantities of water during the cooling process. However, water is still required for other processes, such as coal washing, desulfurization and domestic uses.

The designed water consumption intensities of different processes in a typical coal power plant with either an open-loop cooling system or a closed-loop cooling system in southern China are detailed in Table 2.1. It can be seen that cooling water makes up the largest proportion (80%) of water consumption in power plants with closed-loop cooling systems, which is followed by wet desulfurization, ash removal, boiler makeup, coal transport and domestic use, in descending order.

The choice of cooling technology also affects water use efficiencies at coal power plants by affecting their energy conversion efficiency. The conversion efficiency of current coal power plants from chemical energy stored in coal to electricity is about 30 to 40%. Open-loop cooling systems use running water and therefore have the highest cooling efficiency, while air-cooling systems have the lowest. High cooling efficiency reduces the backward pressure exiting steam exacts on the turbine and therefore improves energy conversion efficiency of the coal power plant. Therefore, power plants with open-loop cooling systems consume less water in other

Table 2.1 Designed water consumption factor of power plants with different cooling systems and capacity

Cooling technology	*Capacity (MW)*	*Wet desulfurization (m^3/h)*	*Dry ash/lime-ash removal (m^3/h)*	*Coal transport (m^3/h)*	*Boiler makeup (m^3/h)*	*Domestic water use (m^3/h)*	*Cooling tower makeup (m^3/h)*	*Total (m^3/h)*	*m3/MWh*
Open-loop	2×300	100–110	109–195	46–83	80–150	5–10		340–548	0.567–0.913
	2×600	190–230	155–247	60–107	90–170	5–10		500–764	0.416–0.637
	2×1000	270–320	265–380	77–146	150–270	5–12		767–1128	0.384–0.564
Closed-loop	2×600	190–230	155–247	60–107	90–170	5–10	2092–2692	2592–3456	2.16–2.88

Source: East China Electric Power Design Institute (2012).

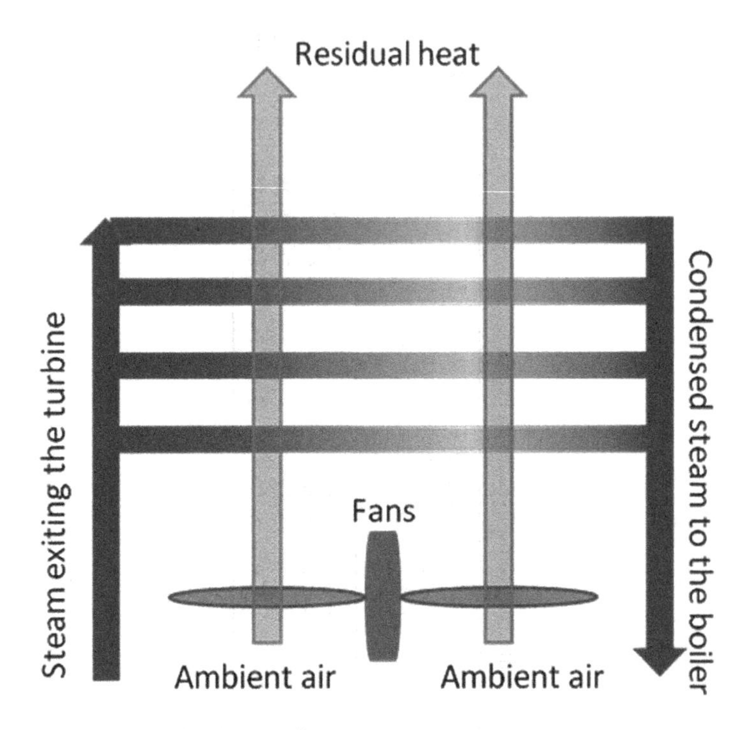

Figure 2.4 Schematic of air-cooling systems

processes due to its higher efficiency. Several hundred coal power plants in China were required to report their water consumption factors to China Electricity Council (CEC, 2012, 2013). A summary of their reported values is presented in Table 2.2.

It can be seen that, for small-scale units, closed-loop cooling systems consume the most water, while air-cooling systems consume the least. However, when the capacity exceeds 600 MW, air-cooling systems lose their advantage to open-loop systems in terms of water consumption. The reason is presumably because air-cooling systems are less efficient and, although they do not need water for cooling purposes, they require a larger amount of coal inputs and hence a larger amount of water used for other purposes, such as coal washing and transporting, wet desulfurization, dust control and so forth.

Reducing water use by deploying air-cooling systems comes at a price. First of all, air-cooling systems require a larger land area, and therefore the capital cost of power plants with dry cooling systems can be at around 2.5 times of that of plants with wet cooling systems (Zhai and Rubin, 2010).

Table 2.2 Average water consumption factors of units of different capacities with different cooling systems in China

Capacity (MW)	*Cooling system*	*Water consumption factor*
<300	Closed-loop	2.25
<300	Open-loop	1.23
<300	Air-cooling	0.47
300–600	Closed-loop	1.93
300–600	Open-loop	0.52
300–600	Air-cooling	0.41
>600	Closed-loop	1.79
>600	Open-loop	0.29
>600	Air-cooling	0.31

Secondly, for the reasons discussed earlier, power plants with air-cooling systems face 5–10% of thermal efficiency losses (Electric Power Research Institute, 2004). Not only does this add to the cost of coal inputs, but it also implies greater greenhouse gas emissions. Consequently, the deployment of air-cooled power plants in China contributed to an additional 24.3–31.9 million tons of CO_2 emissions in 2012 (Zhang et al., 2014). In summary, the choice of cooling technologies should be made based on a comprehensive evaluation considering multiple factors, including local water availabilities, coal prices, power plant construction costs, climate change mitigation and so forth.

Table 2.2 also shows that water use intensities differ by boiler size. The reason is because larger units tend to adopt more advanced technologies, such as supercritical systems that have higher energy conversion efficiencies and therefore lower water intensities (Liao et al., 2017).

Other generating technologies can also affect coal power plants' water use intensity. Macknick et al. (2012) reviewed a large amount of existing literature and technical documents on water use in American power plants with different fuels and generating technologies and summarized their average operational water use intensity as shown in Figure 2.5.

It can be seen that compared to generic, subcritical and supercritical coal power plants, Integrated Gasification Combined Cycle (IGCC) technology lowers the water use intensity, while carbon capture and storage (CCS) increases water use. IGCC uses a high-pressure gasifier to turn coal into pressurized gas – synthesis gas and then use the synthesis gas as the primary energy to generate electricity. Water consumption is dramatically reduced in an IGCC power plant since the syngas is combusted in a gas turbine, and steam is not used as the primary way to convert the energy from the primary energy source (i.e. coal) to electricity. Therefore, a much lower amount of

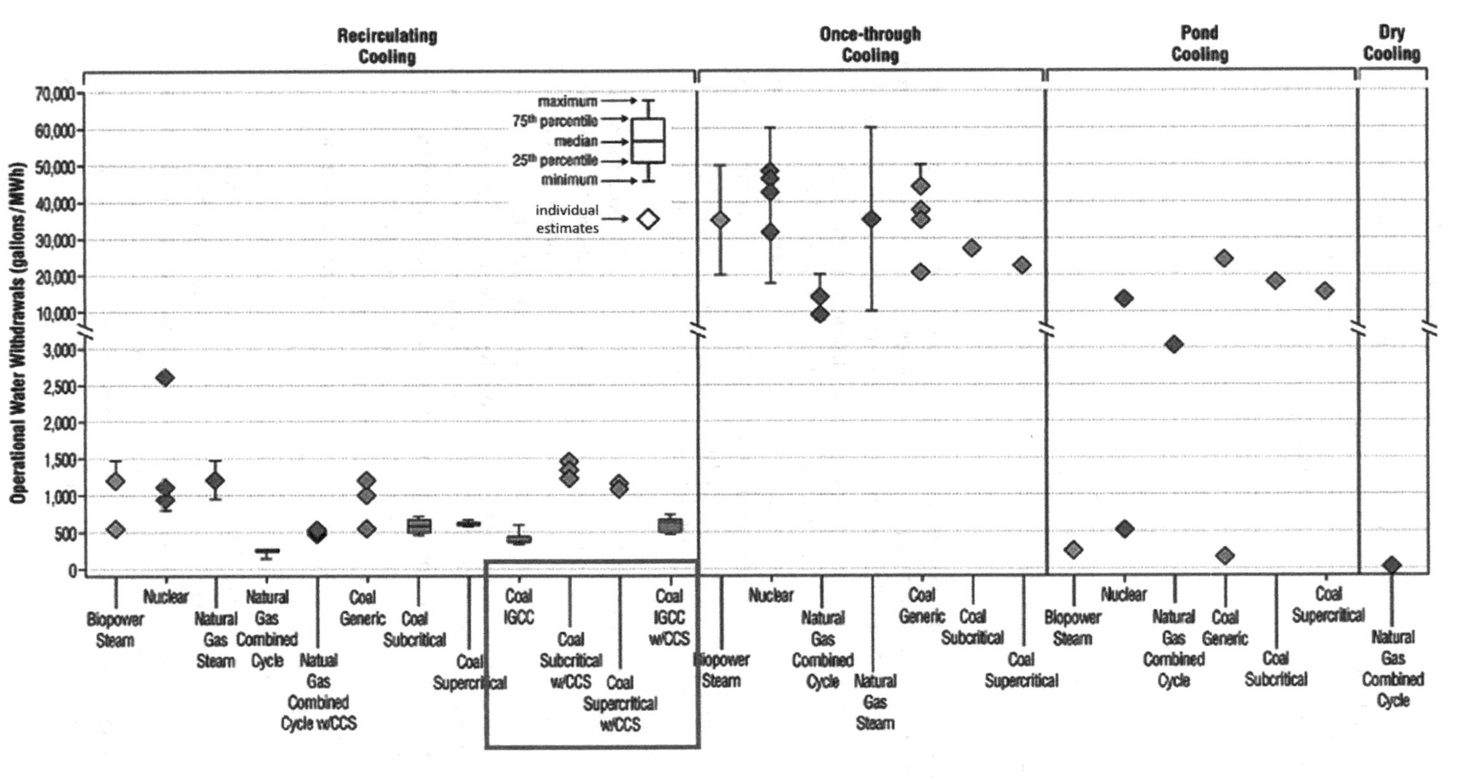

Figure 2.5 Operational water withdrawals for fuel-based electricity-generating technologies. IGCC: Integrated Gasification Combined Cycle. CCS: carbon capture and storage

Source: Adapted from Macknick et al., 2012.

water is needed to condense the steam exiting the turbines as in a normal pulverized coal power plant.

When carbon capture and storage (CCS) is installed at a power plant, an amine-based solvent is usually used to absorb CO_2 from the flue gas at a direct contact cooler (Rochelle, 2009). The flue gas then needs to be washed by water to remove any residual ammonia. The CO_2-rich solvent is then pumped to a stripper where the CO_2 is separated by heat. It should be noted that the heat is often provided by low-quality steam extracted from the steam turbine, which also incurs efficiency loss in the steam cycle. The concentrated CO_2 gas stream is then compressed and transported to a storage site. Water is used for cooling the direct contact cooler, washing flue gas and supporting the CO_2 absorber and stripper. Moreover, as discussed previously, additional water is required for the steam cycle due to the efficiency loss incurred (Zhai, Rubin and Versteeg, 2011). As a result, water intensity almost doubles when CCS is installed to capture emissions from coal power plants (Webster, Donohoo and Palmintier, 2013; Byers et al., 2016).

Ambient and river water temperature

As cooling systems use either air or water as the heat sink, ambient and water temperatures can influence the cooling system's cooling efficiency. The lower the ambient or intake water temperature is, the higher the cooling efficiency gets. In 1824 Nicolas Léonard Sadi Carnot discovered that the maximum energy conversion efficiency of a heat engine is decided by the temperature difference between its hot heat source and cold heat sink, as seen in Figure 2.6. A heat engine refers to a system that converts heat or thermal energy and chemical energy to mechanical energy.

In Figure 2.6, W refers to the work/output of the engine, Q_{HOT} is the heat input at a high temperature, Q_{COLD} is the heat rejected at a low temperature. W equals to Q_{HOT} minus Q_{COLD}. According to Carnot, the ratio between Q_{HOT} and Q_{COLD} equals the ratio between the high-temperature T_{HOT} and low-temperature T_{COLD}. Jiang and Ramaswami (2015) highlighted the seasonal variations of power plants' water intensities based on field data of 19

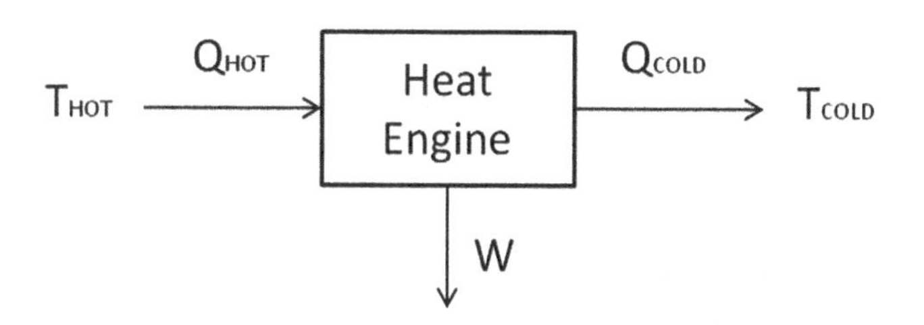

Figure 2.6 A heat engine operating between two temperatures

coal-fired power plants in Shandong, China. Power plants' water intensities are much higher in July when both ambient and river water temperatures are high, and therefore cooling efficiencies are low.

As a result, air-cooling systems' efficiency decreases as ambient air temperature increases, whereas open-loop cooling systems are mostly vulnerable to increased river water temperature, especially when the maximum discharged water temperature is regulated (Forster and Lilliestam, 2009). France and southern California had to curtail their electricity productions in 2003 and 2007, respectively, due to the compounded effects of low flows and increased river water temperatures. Moreover, the warm water discharge from thermoelectric power plants with open-loop cooling systems can potentially cause harmful impacts to river and marine ecosystems by interfering with fish breeding and species range (Raptis and Pfister, 2016; Raptis, Van Vliet and Pfister, 2016; Raptis, Boucher and Pfister, 2017).

Alternative energy sources for electricity

Apart from coal power plants, electricity generated from other primary energy also requires water inputs at different stages. According to Macknick et al. (2012), apart from cooling technologies, fuel types also have significant impacts on power production's water uses, mostly due to the differences in thermal efficiencies. To generate a certain amount of electricity, gas-fired power requires the least water while nuclear power needs the most. Gas-fired power generation requires the least water because all-natural gas plants use a gas turbine where natural gas is mixed with air and combusts and expands to cause a generator to produce electricity. No water is required for generating or cooling the steam in a gas turbine cycle. However, gas turbines can be used in combination with a steam turbine in a combined cycle power plant with very high efficiency. Nuclear technologies require the largest water withdrawals among the thermoelectric generating technologies (EIA, 2019). In order to avoid damages to the finely engineered nuclear fuel assemblies, nuclear power plants operate at a lower temperature than conventional coal or gas power plants and therefore have lower thermal efficiency. Moreover, nuclear power plants do not lose heat through combustion gases; therefore, they withdraw and consume more water per unit of electricity produced (World Nuclear Association, 2013).

In terms of hydropower, although Mekonnen and Hoekstra (2012) have highlighted that hydropower is a significant water consumer because of water evaporation in the dammed reservoirs, there exist many methodological disputes, especially on how to attribute the water consumption to different uses (e.g. hydropower, agriculture, navigation, flood control) in multi-purpose reservoirs (Bakken et al., 2013, 2016). Furthermore, evaporation rates from open water exhibit substantial geographical differences affected by local climatic conditions, such as humidity, wind, temperature and so forth (Scherer and Pfister, 2016; Hogeboom, Knook and Hoekstra, 2018). As a result, the literature estimates exhibit huge uncertainties (Zhang and Anadon, 2013; Bakken, Killingtveit and Alfredsen, 2017). Zhao and Liu (2015) proposed to allocate the water consumption to reservoirs' different uses according to their ecosystem services. Based on this allocation method, Liu et al. (2015) calculated the water consumption intensities for electricity production of China's 209 hydropower plants. The following sections illustrate how water is used for hydropower, gas-fired power and concentrated solar power.

Hydropower

There are four main types of hydropower technologies: **(1) Storage Hydropower Stations (Figure 2.7):** Water stored in a dammed reservoir can be released to transform its potential energy to kinetic energy to drive a water turbine and generate electricity.

(2) Run-of-the-River Hydropower Stations (Figure 2.8): Some hydropower stations have small or no reservoir capacity so that only the water coming from upstream can be used for generation at that moment. The kinetic energy in running water is converted to mechanical energy at a turbine and spins the generator to produce electric energy.

(3) Pumped storage hydropower stations (Figure 2.9) use excess electricity when electricity demand is lower than production to pump water to higher reservoirs acting as a type of energy storage and release water when electricity demand is high. When electricity demand is low, excessive electricity is used to pump water in the lower water body to the upper reservoir. Excessive electric power is converted to potential energy stored

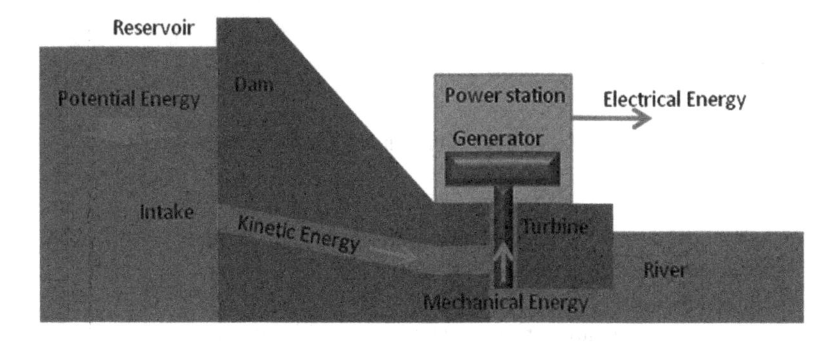

Figure 2.7 Schematic of a hydroelectric power station

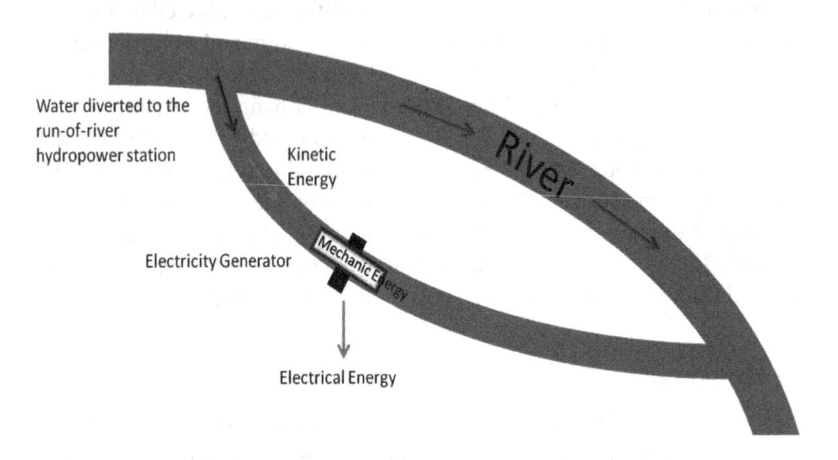

Figure 2.8 A typical run-of-river hydropower station

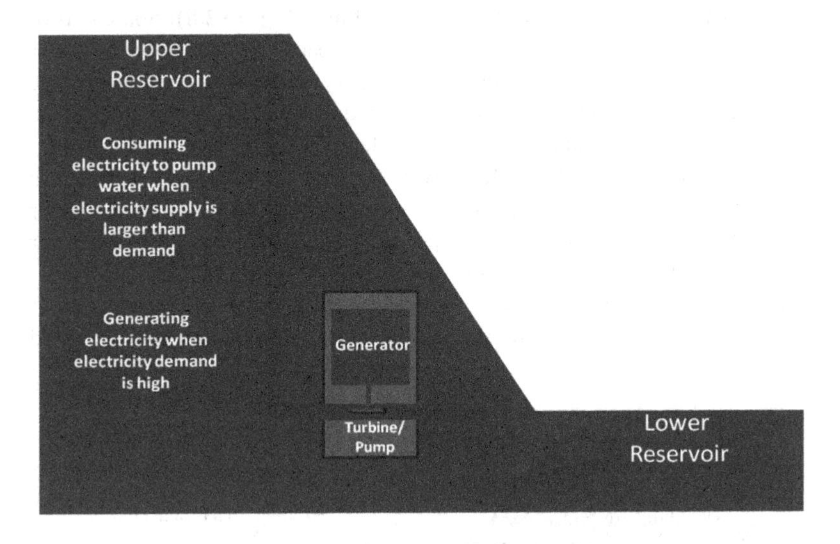

Figure 2.9 Schematic of a pumped storage hydropower station

in the upper reservoir. During the high-electricity demand period, water is released from the upper reservoir to drive the turbine and generator to produce electricity.

In addition, there are various offshore renewable energy technologies, but they are not considered here as they do not consume freshwater.

The capacity of hydropower plants can be calculated according to the following equation:

$$P = m \times g \times Hnet \times \eta$$

P is power (watts); m refers to the water flow (kg s^{-1}); g is the gravitational constant (i.e. 9.81m s^{-2}); Hnet is the net head and η is the conversion efficiency of hydroelectric power plants, which is the product of efficiencies of all different components, including the turbine, drive system and generator. Modern hydropower stations have an energy conversion efficiency of about 90%, while smaller plants may have efficiencies between 60%–80% (Nazari-Heris and Mohammadi-Ivatloo, 2017). Net head is the gross head minus head losses. Gross head refers to the vertical distance between the water intake and the turbine, which can be measured on site. However, there are head losses due to pipeline friction, and net head is calculated as the gross head that can be measured at the site, deducting any head losses. A properly designed pipeline normally yields a net head of about 85%–90% of the gross head (Canyon Hydro, 2013).

Natural gas electric power plants

There are primarily three types of natural gas-fired electric power plants: natural gas combined cycle (NGCC) power plants, gas combustion turbines and gas-fired steam turbines. While gas turbines are directly driven by hot combustion gases, an NGCC plant first uses a gas combustion turbine to generate electricity and then uses the waste heat to create steam to generate more electricity in a steam turbine. Because gas combustion turbines require no cooling, as there is no steam to be condensed, the overall combined cycle system requires much less water for cooling than traditional steam turbine technologies. According to Macknick et al. (2012), water use values (median) of natural gas-fired power plants are shown in Table 2.3.

According to the US Energy Information Administration (2012), more than 80% of natural gas-fired power plants in the US are NGCC plants. Gas turbine and steam turbine power plants account for less than 10% each.

Table 2.3 Water use at natural gas-fired power plants

	Once-through		*Closed-loop*		*Air-cooling*	
	Withdrawal	*Consumption*	*Withdrawal*	*Consumption*	*Withdrawal*	*Consumption*
NGCC	11380	100	253	198	2	2
Steam	35000	240	1203	826	2	2

Source: Macknick et al., 2012.

Concentrated solar power (CSP) plants

Concentrated solar power (CSP) requires larger amounts of water at the operational stage compared to other commonly used renewable energies (i.e. solar PV and wind). Both CSP and solar photovoltaic (solar PV) require water to clean the mirrors and panels, as dust can reduce the system efficiency. Water intensity for mirror and panel washing ranges from 0.08 to 0.15 m^3 MWh^{-1} (Bracken et al., 2015). Such variation depends on the frequency of cleaning, which is decided by the site conditions, such as soil and dust properties, vegetation, air pollution, wind speed and direction, humidity and temperature, as well as precipitation characteristics (i.e. intensity, frequency and duration). Other factors are panel/mirrors orientation and angle of tilt; glazing properties also have impacts on the cleaning frequency (Sarver, Al-Qaraghuli and Kazmerski, 2013).

Apart from cleaning needs, the thermal cycle of parabolic trough, linear Fresnel and power tower CSP technologies are essentially the same as those used in coal and nuclear power plants and therefore require water for cooling purposes as well as boiler makeup, as in a coal-fired power plant.

Dish systems do not generally require water for cooling, nor for steam cycle operations, but do require a small amount to wash the concentrators.

References

AQUASTAT. (1998). *AQUASTAT definitions*. Rome, Italy: FAO.

Bakken, T. H., Killingtveit, A. and Alfredsen, K. (2017). The water footprint of hydropower production – state of the art and methodological challenges. *Global Challenges*, 1, p. 1600018.

Bakken, T. H., Killingtveit, A., Engeland, K., Alfredsen, K. and Harby, A. (2013). Water consumption from hydropower plants – review of published estimates and an assessment of the concept. *Hydrology and Earth System Sciences*, 17, pp. 3983–4000.

Bakken, T. H., Modahl, I. S., Raadal, H. L., Bustos, A. A. and Arnoy, S. (2016). Allocation of water consumption in multipurpose reservoirs. *Water Policy*, 18(4), pp. 1–17.

Bracken, N., Macknick, J., Tovar-Hastings, A., Komor, P., Gerritsen, M. and Mehta, S. (2015). *Concentrating solar power and water issues in the US Southwest*. Golden, CO: National Renewable Energy Laboratory (NREL).

Byers, E. A., Hall, J. W., Amezaga, J. M., O'Donnell, G. M. and Leathard, A. (2016). Water and climate risks to power generation with carbon capture and storage. *Environmental Research Letters*, 11, p. 024011.

Canyon Hydro. (2013). *Guide to hydropower: An introduction to hydropower concepts and planning*. Deming: Canyon Hydro.

China Electricity Council (CEC). (2012). *Notification of energy efficiency benchmarking and competition data of 2012 national 600MW thermal power units* (in Chinese). Beijing, China: CEC.

China Electricity Council (CEC). (2013). *Notification of energy efficiency benchmarking and competition data of 2013 national 300MW thermal power units* (in Chinese). Beijing, China: CEC.

East China Electric Power Design Institute. (2012). Investigation on designed water withdrawal and water consumption statistics of Shanghais big thermal power plants. (Internal materials). Shanghai, China.

Electric Power Research Institute (EPRI). (2004). *Comparison of alternate cooling technologies for U.S. power plants: Economic, environmental, and other tradeoffs*. Palo Alto, CA: EPRI.

Energy Information Administration (EIA). (2012). *Annual electric utility data*. Washington, DC: EIA.

Energy Information Administration (EIA). (2019). *Average tested heat rates by prime mover and energy source, 2008–2018*. Washington, DC: EIA.

Forster, H. and Lilliestam, J. (2009). Modelling thermoelectric power generation in view of climate change. *Regional Environmental Change*, 10(4), pp. 327–338.

Hogeboom, R., Knook, L. and Hoekstra, A. Y. (2018). The blue water footprint of the world's artificial reservoirs for hydroelectricity, irrigation, residential and industrial water supply, flood protection, fishing and recreation. *Advances in Water Resources*, 113, pp. 285–294.

Jiang, D. and Ramaswami, A. (2015). The 'thirsty' water-electricity nexus: Field data on the scale and seasonality of thermoelectric power generation's water intensity in China. *Environmental Research Letters*, 10, p. 024015.

Liao, X. W., Hall, J. W. and Eyre, N. (2017). Water for energy in China. In: L. M. Pereira et al., eds., *Food, energy and water sustainability: Emergent governance strategies*. London: Earthscan, Routledge.

Liu, J., Zhao, D., Gerbens-Leenes, P. W. and Guan, D. (2015). China's rising hydropower demand challenges water sector. *Scientific Reports*, 5, p. 11446.

Macknick, J., Sattler, S., Averyt, K., Clemmer, S. and Rogers, J. (2012). The water implications of generating electricity: Water use across the United States based on different electricity pathways through 2050. *Environmental Research Letters*, 7, p. 045803.

Mekonnen, M. M. and Hoekstra, A. Y. (2012). The blue water footprint of electricity from hydropower. *Hydrology and Earth System Sciences*, 16, pp. 179–187.

Nazari-Heris, M. and Mohammadi-Ivatloo, B. (2017). Design of small hydro generation systems. In: G. B. Gharehpetian and S. Mohammad Mousavi Agah, eds., *Distributed generation systems – design, operation and grid integration*. Oxford: Butterworth-Heinemann.

Raptis, C. E., Boucher, J. M. and Pfister, S. (2017). Assessing the environmental impacts of freshwater thermal pollution from global power generation in LCA. *Science of the Total Environment*, 580, pp. 1014–1026.

Raptis, C. E. and Pfister, S. (2016). Global freshwater thermal emissions from steam-electric power plants with once-through cooling systems. *Energy*, 97, pp. 46–57.

Raptis, C. E., Van Vliet, M. T. H. and Pfister, S. (2016). Global thermal pollution of rivers from thermoelectric power plants. *Environmental Research Letters*, 11, p. 104011.

Rochelle, G. T. (2009). Amine scrubbing for CO_2 capture. *Science*, 325(5948), pp. 1652–1654.

Sarver, T., Al-Qaraghuli, A. and Kazmerski, L. L. (2013). A comprehensive review of the impact of dust on the use of solar energy: History, investigations, results, literature, and mitigation approaches. *Renewable and Sustainable Energy Reviews*, 22, pp. 698–733.

Scherer, L. and Pfister, S. (2016). Global water footprint assessment of hydropower. *Renewable Energy*, 99, pp. 711–720.

Stone, J. C., Singleton, F. D. Jr, Gadkowski, M., Salewicz, A. and Sikorski, W. (1982). Water demand for generating electricity: a mathematical programming approach with application in Poland. International Institute for Applied Systems Analysis, Laxenburg, Austria.

Webster, M., Donohoo, P. and Palmintier, B. (2013). Water-CO2 trade-offs in electricity generation planning. *Nature Climate Change*, 3, pp. 1029–1032.

World Nuclear Association. (2013). *Cooling power plants*. London. Available at: www.world-nuclear.org/information-library/current-and-future-generation/cooling-power-plants.aspx#.Udx22awzZqM.

World Resource Institute. (2015). *Opportunities to reduce water use and greenhouse gas emissions in the Chinese power sector*. Beijing, China: World Resource Institute.

Zhai, H. and Rubin, E. S. (2010). Performance and cost of wet and dry cooling systems for pulverized coal power plants with and without carbon capture and storage. *Energy Policy*, 38(10), pp. 5653–5660.

Zhai, H., Rubin, E. S. and Versteeg P. L. (2011). Water use at pulverized coal power plants with post-combustion carbon capture and storage. *Environmental Science & Technology*, 45, pp. 2479–2485.

Zhang, C. and Anadon, L. D. (2013). Life cycle water use of energy production and its environmental impacts in China. *Environmental Science & Technology*, 47, pp. 14459–14467.

Zhang, C., Anadon, L. D., Mo, H., Zhao, Z. and Liu, Z. (2014). Water-carbon trade-off in China's coal power industry. *Environmental Science & Technology*, 48, pp. 11082–11089.

Zhao, D. and Liu, J. (2015). A new approach to assessing the water footprint of hydroelectricity power based on allocation of water footprints among reservoir ecosystem services. *Physics and Chemistry of the Earth*, pp. 40–46, 79–82.

3 Water use in coal power plants' upstream fuel cycle

Development of China's coal industry

China is the world's largest producer and consumer of coal. Coal production in China roughly accounts for 45% of the global total, and coal meets about 60% of China's primary energy consumption, which is far higher than that in OECD countries, at 20%. China's economic development has relied to a large extent on the utilization of coal. From 1980, since China started its 'Reform and Opening Up' policy and its rapid economic development, its coal production has increased in the same period and reached the maximum of 3.97 giga ton (Gt) in 2013. Its major coal-producing provinces include Inner Mongolia, Shanxi and Shaanxi, mostly located in the north. Take Shanxi, for example, while it is the second-largest coal-producing province in China, following Inner Mongolia, its production, 893 million tons in 2018, surpasses India, the second-largest coal producer in the world.

In terms of coal consumption, electricity generation accounts for about 50% of coal consumption in China. In 2013, China's coal consumption accounted for more than 50% of the global consumption, amounting to 4.2 and 6.7 times that of the US and European Union (BP, 2014).

This enormous coal use has had a series of environmental impacts, especially on the environment. China overtook the US in 2006 as the world's biggest CO_2 emitter and currently accounts for 30% of the global emissions, while coal is the largest source for CO_2 emissions in China, contributing to 70% of the national total (Shan et al., 2018).

The combustion of coal is also the largest source of air pollutants of particulates, SO_2, and NO_x in China and contributes to China's worsening air pollution problems, which are the fourth highest causes of death and morbidity in China and consequently has drawn much public attention. Researchers from Peking University, Tsinghua University, MIT and other institutions have found that burning coal in China's northern regions has greatly increased total suspended particulates (TSPs) air pollution and caused 500 million residents to lose more than 2.5 billion life years of life expectancy (Chen et al., 2013). Furthermore, the ill-managed coal mining industry has also been accused of many disastrous incidents and the chronic mortality rate. For example, there were 375 coal mining-related deaths in 2017 alone in China, according to its National Coal Mine Safety Administration.

Recognizing the urgency of global climate change mitigation and tackling air pollution problems, the Chinese government has undertaken various policy measures to curb the expansion of its coal industry. China's Intended Nationally Determined Contributions (INDC) have set targets for increasing the share of non-fossil fuels in its primary energy portfolio to over 20% by 2030 (National Development and Reform Commission, 2015), which corresponds to significant reductions in its coal uses. In its 13th Five-Year Plan, the government of China also adopted a coal consumption cap of 4.1 billion tons and a total energy consumption cap of 5 billion tons of coal equivalent (tce) for 2020 (National People's Congress of China, 2016). After peaking at 4.24 billion tons in 2013, China's coal demand has kept decreasing to 3.85 billion tons in 2016 and then slightly bounced back to 3.91 billion tons in 2018, which is still well within the cap set for 2020. The share of coal consumption in the primary energy mix was also significantly reduced from 66% in 2013 to 59% in 2018.

Water use in coal power plants' upstream fuel cycle

Water use for coal-fired power plants' upstream fuel cycle is mostly related to direct extraction and processing of coal (Gleick, 1994, Figure 3.1), while coal transportation also induces water usage due to the use of slurry. Coal

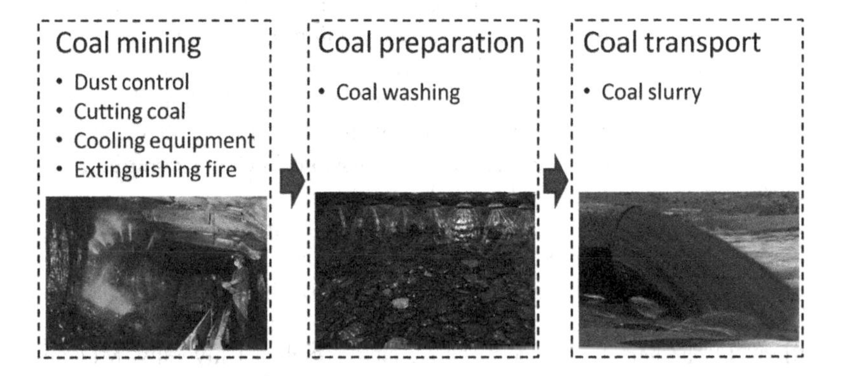

Figure 3.1 Summary schematic of water use in coal power plants' upstream fuel cycle

slurry is a mixture of crushed coal suspended in liquid, usually water, as a means of transporting coal.

According to the US Department of Energy (DOE, 2006), daily water use for coal mines, including coal washing, amounts to 0.26 to 0.98 million m^3 in the US. Water is used in coal mining for cutting coal, dust control, cooling equipment, washing, extinguishing fire and other activities. The amount needed depends on the characteristics of the mine (surface or underground), as well as processing and transport requirements. While about 95% of coal is produced from underground coal mines in China (Meng et al., 2009), they normally require larger water inputs than surface mining. According to the US Department of Energy (DOE, 1983), underground coal mining requires more water for dust control inside the coal mines, which accounts for about 70% of the total on-site mining water use, while the rest is used for coal washing. Moreover, underground mining also requires more upstream water use than surface mining because of the extensive use of mine equipment that requires water inputs in their manufacturing stages. However, the water consumed in the upstream manufacturing stage is not considered in this book. There are substantial variations in the exact values of water uses, depending on sources of the data, and conditions, such as heating value and seam thickness of the fuel (Fthenakis and Kim, 2010). Smaller coal mines generally have higher water intensities. Moreover, for coal mine safety considerations, a large amount of groundwater is pumped in underground coal mines and drained to the surface. It is estimated that China's annual coal mine drainage can reach 4.2 km^3 (Meng et al., 2009). Overall, due to mining and draining activities, groundwater aquifers can be severely damaged, which leads to water-level decline, land surface subsidence, land desertification issues and so on. Take Shanxi, for example, where it is estimated that more than 1 m^3 of groundwater is drained to mine one ton of coal (Wang, 2010). Besides water quantity depletion, water quality of both surface water and groundwater risks being heavily polluted by mining activities (Meng et al., 2009; Bian et al., 2010). Coal mine water is typically contaminated with metals, including iron, copper, magnesium and nickel (Tiwary, 2001). Depending on the geology, the mine water may be highly acidic. Because of the large quantities of mine water that can be produced, water purification and safe disposal of contaminants can be problematic.

The next water-requiring stage in the fuel cycle is coal preparation. Coal preparation includes processes to remove noxious minerals from raw coal (Shan and Ye, 1999). Coal washing is an important method to improve coal quality but also requires water as an input medium. The amount of water required depends on the design of the plant, number of washing stages and other factors. Coal of better quality, after being washed, has higher combustion efficiency and therefore a higher wash rate is desired.

However, the coal wash rate is relatively low in China. In order to increase its coal wash rate, China has been active in constructing a large number of coal preparation plants. By the end of 2007, 1,300 coal preparation plants of a total production capacity of 1.25 billion tons per year have been operating in China (Wu, 2009). However, the coal washing rate was still only less than 45% until 2009, much less than other coal-rich countries, where more than 55% of their coal is washed (Pan et al., 2012).

There are two types of methods for coal preparation, wet methods and dry methods. While more than 90% of coal preparation plants in China use the traditional wet methods, dry methods offer a number of significant advantages. First of all, dry methods significantly lower the water requirements and therefore can be promoted in water-scarce regions. On average, 2.5 m^3 of water are withdrawn to prepare one ton of raw coal (Zhou, 2005). However, because the majority of the water withdrawn will be recycled in a closed loop, only a small fraction is consumed. The average water consumption intensity is 0.2 m^3 per ton of coal, of which 0.15 m^3 remains in the products and the other 0.05 m^3 is consumed in the preparation process (Zhong, 2001). Second, surface moisture gain after wet cleaning can offset the improvement in energy content derived from ash reduction. Third, some low-rank coals can be broken down upon exposure to water, generating excessive fines and resulting in moisture and handling problems. Since a large proportion of coal produced in China is young and can easily degrade in water, dry preparation is becoming a more attractive option.

In the final stage of the fuel cycle, slurry pipelines for coal transportation also require a large amount of water, whereas train transportation is much more commonly used and requires negligible amount of water inputs.

There have been a number of studies, both internationally and domestically, looking at the water withdrawal or consumption intensities for the coal power sector's upstream fuel cycle, including coal mining, preparation and transportation. There are mainly two methods that can be used for such purposes: bottom-up process-based lifecycle inventory analysis and top-down environmentally extended input-output analysis. With the bottom-up lifecycle inventory analysis, Mekonnen, Gerbens-Leenes and Hoekstra (2015) estimated that 340 km^3 of water is consumed every year for fuel extraction, power plant construction and power plant operation of the global electric power generation's lifecycle. Meldrum et al. (2013) reviewed the literature estimates in the US of both water withdrawal and water consumption for the full life cycle of electricity generation, which includes fuel production and processing. They conducted a comprehensive and systematic review of the literature through a broad search of publicly available sources across a range of different disciplines and publication types and applied the same screens for quality and relevance. The median, maximum and minimum values of water use intensities of the coal mining, processing and transport for different types of coal mines, as listed in Table 3.1.

There are also governmental policies and official requirements on water use for coal mines. According to the Industrial Water Use Standard of Shanxi Province, large-scale underground coal mines can only consume up to 0.25 to 0.30 m^3 of water to mine one ton of coal (Shanxi Provincial Government, 2008). In provinces with various types of water and coal resources, water use standards for one ton of coal mining production range from 0.06 to 1.6 m^3. Based on those water intensity values and coal mining activity data (i.e. the amount of coal mined and prepared), some scholars have estimated the total water use for coal mining, preparation and transportation in China. Liao and Ming (2019) estimated that, in 2015, 16.52 km^3 of water was consumed by energy productions in China, including coal, coke, oil, natural gas and thermoelectricity, among which coal production consumed 9.52 km^3 of water and dominated the total water consumption by energy productions. Among water uses for all energy productions, coal production played a dominant role in terms of water consumption, mostly in traditional coal-producing provinces (i.e. Shanxi (87%), Inner Mongolia (85%), Shaanxi (79%), Guizhou (69%) and Yunnan (61%)). The top five provinces with the highest water consumption by coal production were Shanxi (2.45 km^3), Inner Mongolia (2.34 km^3), Shaanxi (1.35 km^3), Guizhou (0.44 km^3) and Xinjiang (0.40 km^3), which is in line with their provincial coal productions. Pan et al. (2012) shed light on potential water uses for China's coal industry supply chain in the near future, 2020 and 2030, under four different scenarios depending on whether the Chinese government will adopt any water-saving policies or technologies in its coal industry. Under different scenarios, expected water uses for coal mining and preparation are detailed in Table 3.2.

It can be seen from Table 3.2 that without any new policies or technology advancements, water use for coal mining and preparation could exceed 11 and 10 km^3 respectively by 2030. Together, policy instruments and

Table 3.1 Water use intensities of upstream fuel cycle for coal power plants

	Water use intensity (gal MWh-1)		
	Median	*Min*	*Max*
Coal mining (surface)	3	0.1–0.5	13
Coal mining (underground)	27	8	180
Coal mining (not specified)	45	12	120
Processing	18	9	1000
Slurry transport	110	100	410

Source: Meldrum et al., 2013.

Table 3.2 Water use for coal mining and preparation in China (Unit: km³)

Scenarios	*Year*	*Coal mining*	*Coal preparation*
Current policies & technology	2008	6.5	3.5
	2020	9.7	7.9
	2030	11.1	10.9
Current policies & technical progress	2008	6.5	3.5
	2020	6.7	4.7
	2030	4.8	2.2
New policies & current technology	2008	6.5	3.5
	2020	9.0	7.3
	2030	9.2	9.0
New policies & technical progress	2008	6.5	3.5
	2020	6.2	4.4
	2030	4.0	1.8

Source: Pan et al., 2012.

technologies can bring down the numbers to 4 and below 2, down by 63% and 83% respectively.

The challenges at the coal-water nexus are highlighted by the location of an estimated 71% of China's coal resources in four water-scarce arid and semi-arid regions of northern China that account for only 13% of total water resources. Moreover, China plans to build 14 large-scale coal bases during the period of the 13th Five-Year Plan (2016–2020), which are largely located in the north (National People's Congress of China, 2016). According to Shang et al. (2016), in 2012, coal industries in those 14 major coal bases together consumed 4.42 km^3 of water, of which 1.51 was consumed by coal production and washing. According to China's National Economic Census Yearbook in 2008 (the latest National Economic Census Yearbook available), 'Coal Mining and Processing' and 'Thermal Electric and Heating Production' withdrew 1.50 and 3.77 billion m^3 water, amounting to 2.23% and 55.76% of China's total industrial water withdrawals. In northern Shanxi and northern Shaanxi, more than 64% of the water consumption associated with coal industries was consumed for coal mining and preparation. Shanxi province is one of the most water-scarce provinces in China, with total water resources estimated at around 13 km^3, with an estimated per capita water availability of 352.65 cubic meters per year. A country is considered 'water stressed' when its total renewable freshwater resources lie between 1,000 cubic meters and 1,700 cubic meters per person per year and 'water-scarce' countries have an average of less than 1,000 cubic meters of renewable freshwater per person per year (Falkenmark, Lundquist and Widstrand, 1989). Clearly the use of water for the coal industry is exacerbating this chronic water scarcity.

To mitigate water issues of coal development, the government of China issued a 'Water Allocation Plan for the Development of Coal Bases' to reduce water usage, improve water efficiency and reduce wastewater discharges in the coal sector. This part of the policy also requires future large-scale coal projects in water-scarce regions to be developed in partnership with local water authorities. The largest potential water savings in the coal mining industry can be realized by increasing the recycling rate of the coal mine drainage for other uses based on the water quality after treatment. The National Energy Administration's 2015 Action Plan for Clean and Efficient Utilization of Coal (2015–2020) has set targets for mine water reuse (i.e. 95% in water-scarce regions and 75% in water-rich areas). In addition, large-scale coal extraction and characterization of aquifers could also reduce water consumption and pollution of water resources from coal mining. Adopting dry preparation technologies and transporting coal using trains are also viable options to reduce water inputs in the upstream fuel cycle of coal power plants.

References

Bian, Z., Inyang, H., Daniels, J., Otto, F. and Struthers, S. (2010). Environmental issues from coal mining and their solutions. *Mining Science and Technology*, 20, pp. 0215–0223.

BP. (2014). *Global energy statistics*. London: BP.

Chen, Y., Ebenstein, A., Greenstone, M. and Li, H. (2013). Evidence on the impact of sustained exposure to air pollution on life expectancy from China's Huai River policy. *Proceedings of the National Academy of Sciences of the United States of America*. https://doi.org/10.1073/pnas.1300018110.

Falkenmark, M., Lundquist, J. and Widstrand, C. (1989). Macro-scale water scarcity requires micro-scale approaches: Aspects of vulnerability in semi-arid development. *Natural Resources Forum*, 13(4), pp. 258–267.

Fthenakis, V. and Kim, H. C. (2010). Life-cycle uses of water in U.S. electricity generation. *Renewable and Sustainable Energy Reviews*, 14, pp. 2039–2048.

Gleick, P. H. (1994). Water and energy. *Annual Review of Environment and Resources*, 19, pp. 267–299.

Liao, X. W. and Ming, J. (2019). Pressure imposed by energy production on compliance with China's 'three red lines' water policy in water-scarce provinces. *Water Policy*, 21, pp. 38–48.

Mekonnen, M. M., Gerbens-Leenes, P. W. and Hoekstra, A. Y. (2015). The consumptive water footprint of electricity and heat: A global assessment. *Environmental Science: Water Research & Technology*, 1, p. 285.

Meldrum, J., Nettles-Anderson, S., Heath, G. and Macknick, J. (2013). Life cycle water use for electricity generation: A review and harmonization of literature estimates. *Environmental Research Letters*, 8, pp. 015–031.

Meng, L., Feng, Q., Zhou, L., Lu, P. and Meng, Q. (2009). Environmental cumulative effects of coal underground mining. *Procedia Earth and Planetary Science*, 1, pp. 1280–1284.

National Development and Reform Commission of China. (2015). *Intended nationally determined contribution*. Beijing, China: NDRC.

National Energy Administration. (2015). *Action plan for clean and efficient utilization of coal (2015–2020)*. Beijing, China: National Energy Administration.

National People's Congress of China. (2016). *The thirteenth five-year plan*. Beijing, China: National People's Congress of China.

Pan, L., Liu, P., Ma, L. and Li, Z. (2012). A supply chain-based assessment of water issues in the coal industry in China. *Energy Policy*, 48, pp. 93–102.

Shan, Y., Guan, D., Zheng, H., Ou, J., Li, Y., Meng, J., Mi, Z., Liu, Z. and Zhang, Q. (2018). China CO_2 emission accounts 1997–2015. *Scientific Data*, 5, p. 170201.

Shan, Z. and Ye, L. (1999). China coal industry. In: *Encyclopaedia, processing utilization & environmental protection volume*. Beijing, China: China Coal Industry Publishing House.

Shang, Y., Lu, S., Li, X., Hei, P., Lei, X., Gong, J., Liu, J., Zhai, J. and Wang, H. (2016). Balancing development of major coal bases with available water resources in China through 2020. *Applied Energy*, 194, pp. 735–750.

Shanxi Provincial Government. (2008). *Water use standard of Shanxi province*. Shanxi Government. Available at: www.shanxigov.cn/n16/n1203/n1866/n5130/n31265/1010341.html.

Tiwary, R. K. (2001). Environmental impact of coal mining on water regime and its management. *Water, Air, and Soil Pollution*, 132, pp. 185–199.

US Department of Energy (DOE). (1983). *Energy technology characterizations handbook: Environmental pollution and control factors*. Washington, DC: DOE.

US Department of Energy (DOE). (2006). *Energy demands on water resources: Report to congress on the interdependency of energy and water*. Washington, DC: DOE.

Wang, Q. (2010). Evaluation on influences of coal mining for water resources and control measures (in Chinese). *Shanxi Hydro Technics*, 175, pp. 14–16.

Wu, S. (2009). Development of Chinese coal preparation in past 30 years (in Chinese). *Coal Processing & Comprehensive Utilization*, pp. 1–4.

Zhong, Y. (2001). Zero draining of waste water in coal preparation industry (in Chinese). *Industrial Water & Wastewater*, 32, pp. 43–44.

Zhou, C. (2005). Current situation and development of slime water treatment technology in China (in Chinese). *Gansu Science and Technology*, 21, pp. 142–143.

4 Water use in China's coal power plants

China's electricity grids

China's electricity grid is divided into six regional grids (i.e. northeast, northwest, north, east, central and south) (Figure 4.1). China's development is highly concentrated along the coast in the south, east and north grids where electricity consumption is particularly high. Inter-grid electricity transmission is small, amounting to less than 5% by 2015. Most electricity consumption in the three northern grids is produced from thermoelectric power, while larger shares of consumed electricity in the central, eastern and south grids come from hydropower productions. It is noticeable that annual available water resource per capita amounts to only 177.29 m^3 in the north grid, where 'extreme water scarcity' is defined as annual renewable freshwater per person of less than 500 cubic meters (Falkenmark, Lundquist and Widstrand, 1989). Although water resource availability is particularly high in the northwest region on average, the intra-regional difference is huge. Within this region, annual available water resource per capita in the Shaanxi, Gansu and Ningxia provinces is only at 879, 634 and 138 cubic meters, classifying them as water-scarce provinces.

Current water use by China's coal power plants

Liao, Hall and Eyre (2016) consolidated multiple datasets on coal power plants in China, including Carbon Monitoring for Action (Center for Global Development, 2015), World Electric Power Plants Data Base (Utility Data Institute, Platts Energy InfoStore, 2015), Global Coal Plant Tracker (CoalSwarm, 2015) and Enipedia (Davis et al., 2015) and identified their cooling technology from Google imageries following the procedures listed in US Geological Survey (Diehl et al., 2013). Over 2,300 coal-fired EGUs, whose total capacity amounts to 870 GW, about 96.7% of the nation's total in 2015, are included.

Figure 4.2 illustrates the current cooling technologies used in China in 2015, demonstrating the regional popularity of different cooling technologies.

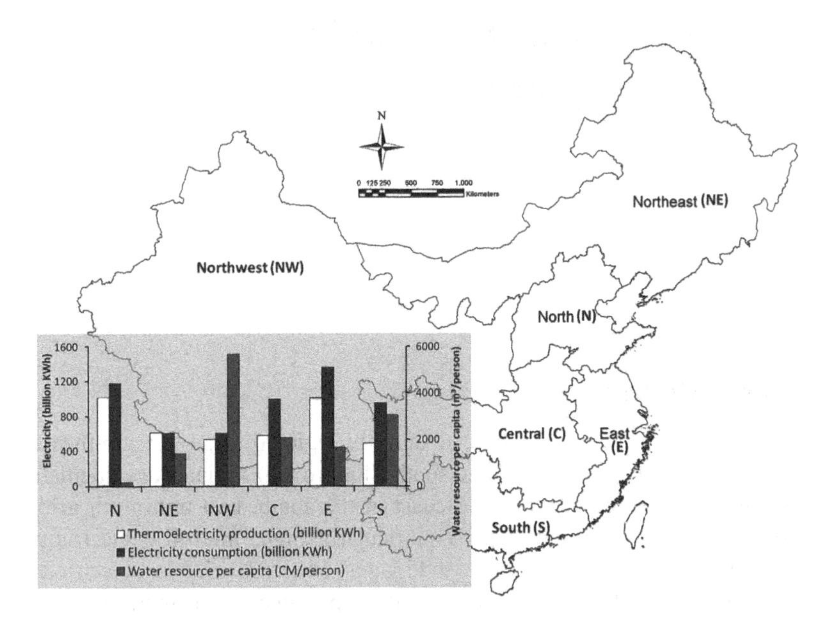

Figure 4.1 Electricity grids in China and their thermoelectricity production, total electricity consumption and annual available water resource per capita

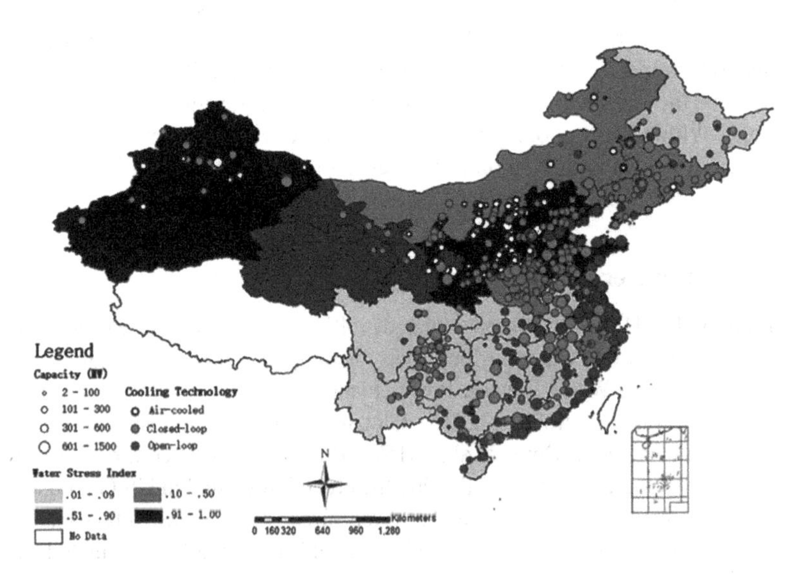

Figure 4.2 China's current coal power plants with different cooling types

Source: Chai et al., 2018.

Closed-loop systems are used throughout the whole country; open-loop systems are commonly adopted in the southeast and coastal regions where water – either freshwater or seawater – is abundant and discharged water does not constitute thermal pollution on the downstream; air cooling is mostly used in northern China where it faces the most severe water shortages.

The provincial configurations are illustrated in Figure 4.3.

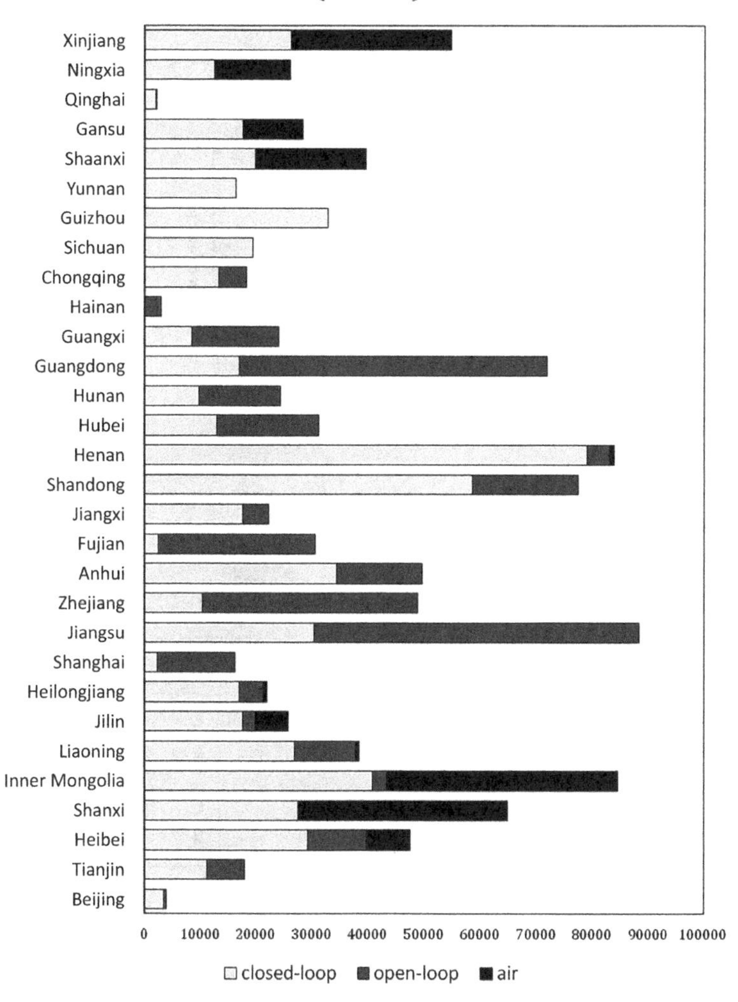

Figure 4.3 Provincial capacity of coal power plants with different cooling technologies

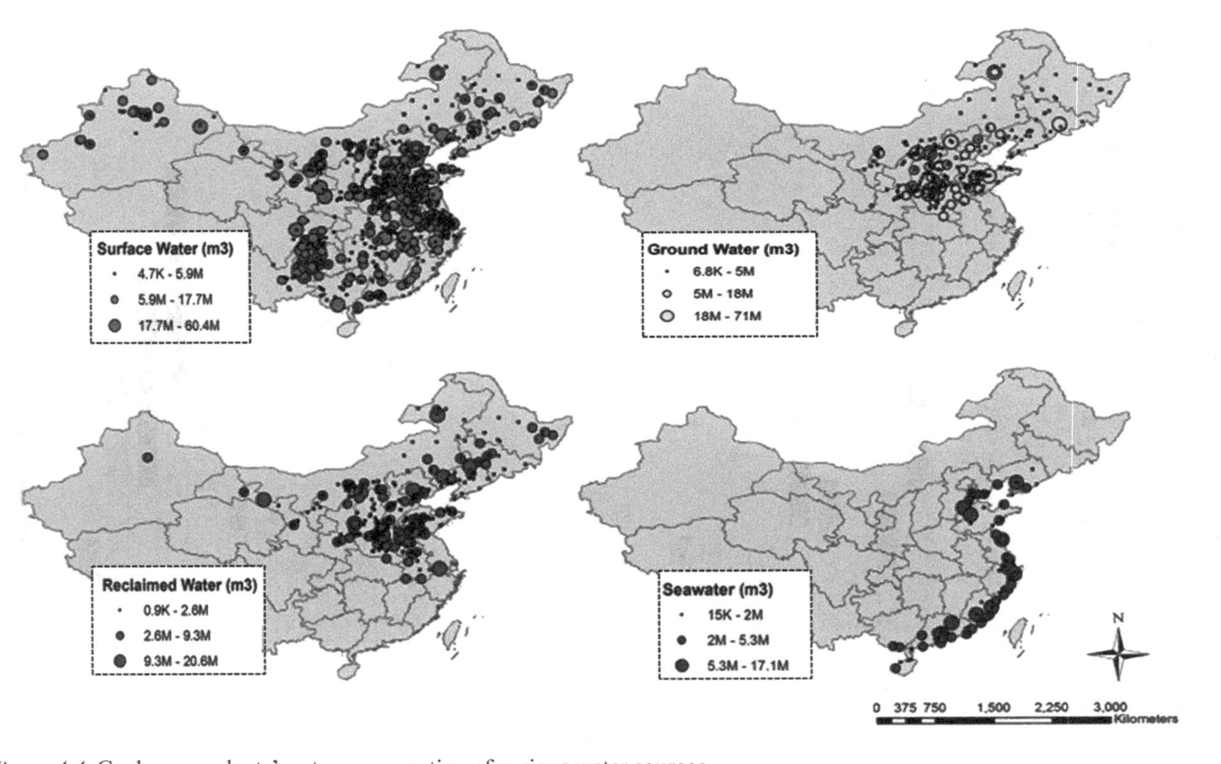

Figure 4.4 Coal power plants' water consumption of various water sources

Source: Liao, Hall and Eyre, 2016.

Water consumption from four sources (surface water, groundwater, reclaimed water and seawater) by power generation is presented in Figure 4.4. In 2015, the national total water consumption and withdrawal (including both freshwater and seawater sources) for power production in China amounted to 4.86 and 124.06 billion m^3. While seawater is unsurprisingly withdrawn along the coast, three hot spots are identified with the largest amounts of surface water consumed for coal power plants, namely Jing-Jin-Ji, Yangtze River Delta and Pearl River Delta megalopolises. These three regions encompass China's fastest-growing mega-cities, such as Beijing, Tianjin, Shanghai, Guangzhou and Shenzhen. An increasingly concentrated population and people's burgeoning lifestyle both require higher levels of electricity provision, which is exerting increasing pressure on the environment and natural resources. Although the government of China issued a ban on groundwater utilization in water-scarce regions in 2004, it is still being heavily used by the existing power plants in the northern China plain (e.g. Hebei, Shanxi, Shaanxi, and Inner Mongolia) together with reclaimed water as supplementary water resources where surface water is lacking.

The pattern of provincial water use is determined by the total electricity consumption, which is highly correlated with population and development status, dependency on thermoelectricity and cooling technology deployment. Water withdrawal is especially high in Jiangsu, Zhejiang and Guangdong due to the high penetration of the open-loop cooling system (Figure 4.5). Water consumption prevails especially in intensively populated provinces with high electricity demands (e.g. Jiangsu, Shandong, Henan) and provinces where closed-cooling technology is predominantly used (e.g. Guizhou, Anhui, Xinjiang).

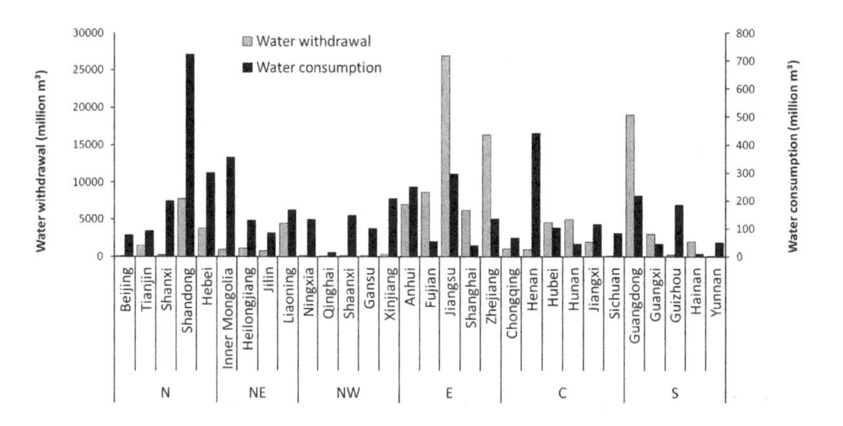

Figure 4.5 Provincial water use by coal power plants

Historical water use by China's coal power plants

From 2000 to 2015, the national total water consumption and withdrawal (including both freshwater and seawater sources) for power production in China grew by more than three times, from 1.25 and 40.75 billion m^3 to 4.86 and 124.06 billion m^3, respectively (Figure 4.6). Water withdrawal peaked in 2013 at 129.82 billion m^3. In 2013, southern China suffered from record-high heat waves that could have contributed to these regions' electricity demand peaking (USA Today, 2013). Open-loop cooling systems are prevalently utilized in the east and south grids where water is abundant (Liao, Hall and Eyre, 2016). As a result, the largest amount of water was withdrawn in the east, from 18.28 billion m^3 in 2000 to 65.01 billion m^3 in 2015 and peaking at 68.16 billion m^3 in 2013, which is followed by the south. Much smaller amounts of water are withdrawn in the northern grids, particularly the northwest, by coal power plants.

The largest amount of water was consistently consumed in the north, from 434.13 million m^3 in 2000 to 1.39 billion m^3 in 2015, imposing increasing pressure on the region's water scarcity. As can be seen from Figure 4.1, China's north grid suffers from the most severe water scarcity. On the other hand, it also houses one of the three rapidly expanding megalopolises (i.e. Jing-Jin-Ji region) and serves as the political center of China, where the capital city Beijing is located. Increasing water demands have led to the over-exploitation of groundwater resources in this region and are causing severe and irreversible ecological and environmental problems, such as biodiversity degradation, groundwater pollution, land surface subsidence and so forth.

Drivers of water use increase in China's coal power plants

According to the Impact=Population x Affluence x Technology (IPAT) model (Ehrlich and Holdren, 1971; Mi et al., 2017), Liao and Hall (2018) attributed China's historical water-for-electricity change to three corresponding factors: (1) population, (2) power production per capita (MWh/p) and (3) water intensities (m^3/MWh). Besides energy sources and cooling technologies, power plants' water intensities are also affected by other technological factors, for example, boiler technology, but their effects are much smaller by an order of magnitude for water consumption and three orders of magnitude for water withdrawal (Macknick et al., 2012).

Effects of population and power production per capita

The increases of population and power production per capita both have had positive effects on the power sector's water use increases (Figure 4.7). In

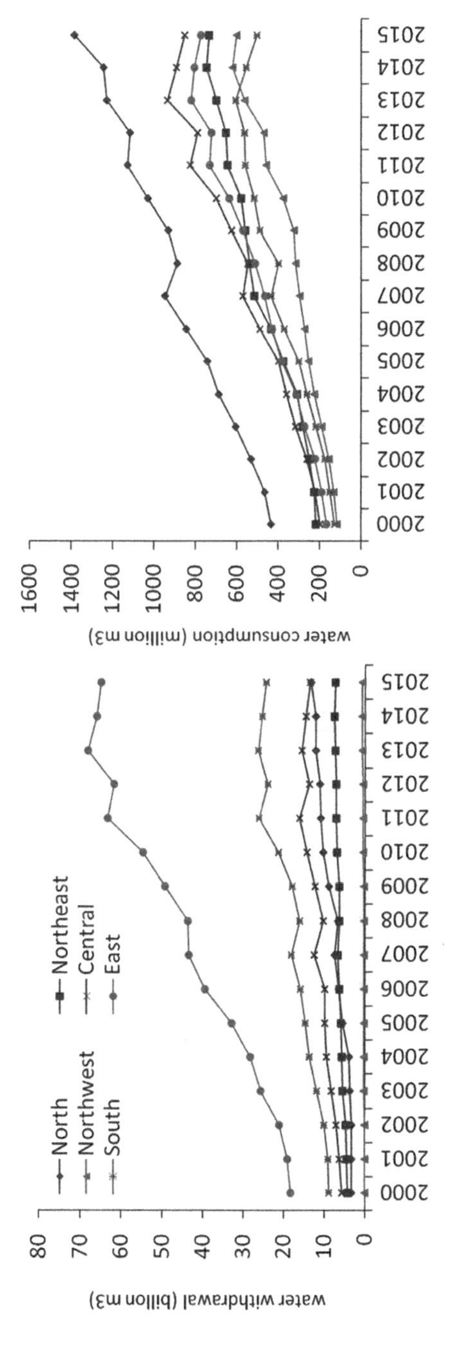

Figure 4.6 China's electric power sector's water withdrawal (left) and consumption (right) from 2000 to 2015

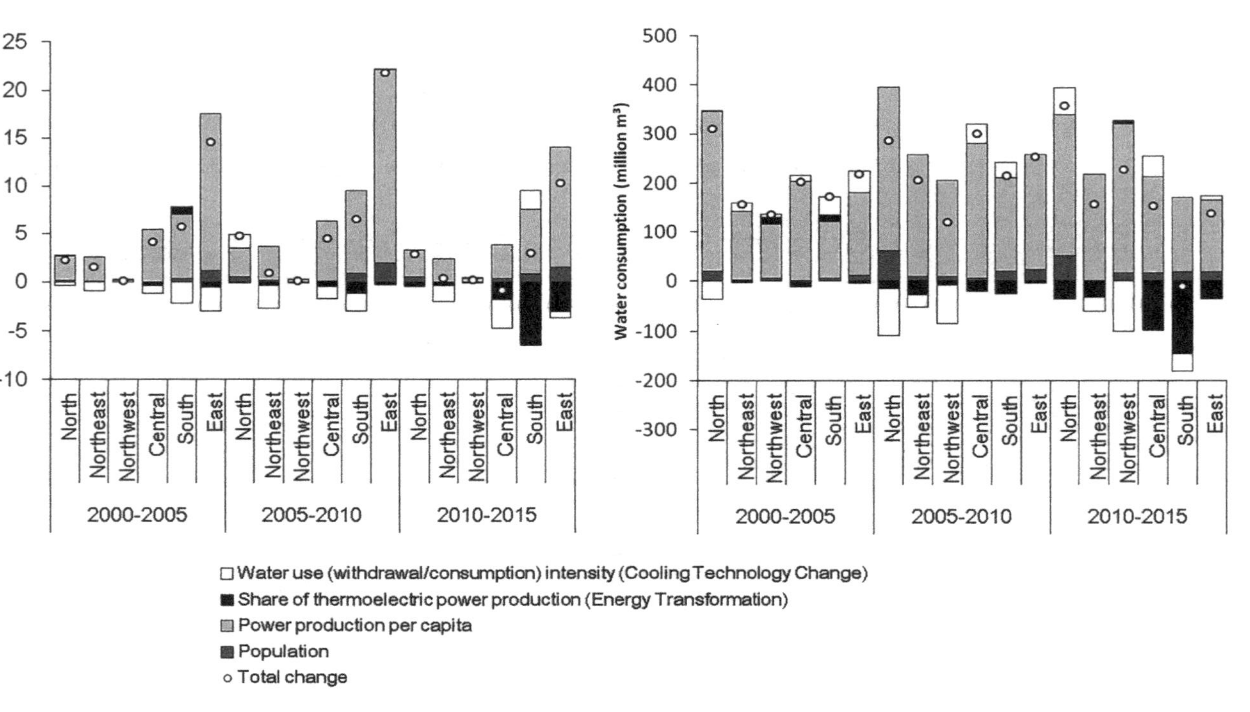

Figure 4.7 Drivers of China's electric power sector's water use changes from 2000 to 2015

Source: Liao and Hall, 2018.

total, from 2000 to 2015, population increase contributed to 8.43 billion m^3 and 295.99 million m^3 of water withdrawal and consumption increases, respectively, compared with 103.40 and 3.84 billion m^3 triggered by the rise of power production per capita. Population played the least important role, which is not surprising considering China's population has stabilized during this period at an average annual growth rate of 0.56%, while power production per capita has kept an annual growth rate of around 10% during the same period. In the east, population growth had made a visible contribution to the water withdrawal increases, whereas the population effect on water consumption was the highest in the north. The north and the east are homes to China's two rapidly expanding urban agglomerations, Jing-Jin-Ji and the Yangtze Delta region respectively. Both regions have attracted significant numbers of inward migrations for better jobs and opportunities (Wang, Wu and Li, 2017), which has resulted in heightened pressure on its local resources.

Growth in power production per capita played the dominant role throughout the whole country on both water withdrawal and consumption. However, China's national average power production per capita is still relatively low, at only 4,153 KWh in 2014 and ranked around 70th in the world, compared to 7,995 KWh in OECD countries and 12,987 KWh in the US for instance (World Bank, 2017).

Effects of energy transformation

As the world's biggest CO_2 emitter, China pledged to increase the share of its non-fossil fuel to around 15% by 2020 at Copenhagen in 2009. It first proposed to diversify the energy portfolio in its 10th Five-Year Plan (2001–2005) (National People's Congress of China, 2001), and the 12th Five-Year Plan (2011–2015) (National People's Congress of China, 2011) set more specific targets, aiming to decrease the share of coal in the country's energy structure to 63%. Among the non-hydropower portfolio, during the study period (2000–2015), even though the sheer amount of electricity produced from fossil fuel has increased drastically from 1088.48 to 4241.93 TWh, the percentage for which it accounts has gone down from 98.5% to just over 91.4%, which was gradually taken over by nuclear, wind power and solar PV.

Such energy transition – the change of the power sector's energy portfolio (Figure 4.8) – has mitigated 14.46 billion m^3 of water withdrawal increase and 429.30 million m^3 of water consumption increase in the power sector from 2000 to 2015. The effect was especially significant in the last five-year period, contributing to 83.67% and 80.00% of the total, respectively.

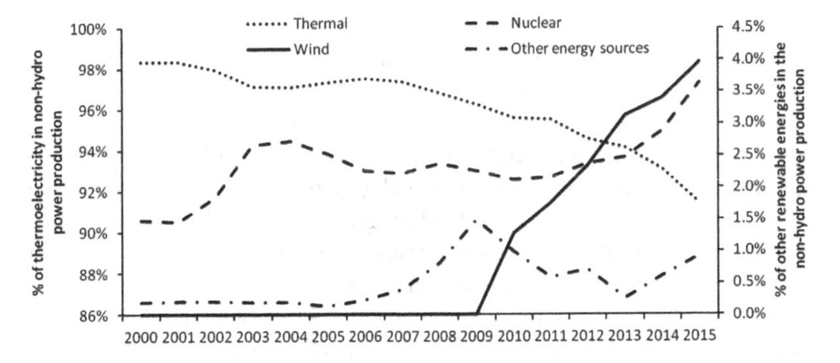

Figure 4.8 Energy structure of China's electric power productions (excl. hydro) from 2000 to 2015 (Note: Wind power and solar PV were included in 'Other' before 2005.)

Source: China Electric Statistic Yearbook, 2015.

The coal power sector is the major source of air pollution and greenhouse gas emissions in China due to its intensive coal combustion. Facing pressing local health impacts and global climate change pressures, China has taken actions to reduce its power sector's reliance on coal by either transforming its energy portfolio or improving coal-burning efficiencies. Such policies can generate co-benefits of water savings. For example, Webster, Donohoo and Palmintier (2013) found that a restriction on CO_2 emissions in the energy sector would also reduce water withdrawals by the power sector in Texas; Bartos and Chester (2014) have shown that Arizona's Energy Efficiency Mandate and Renewable Portfolio Standard have resulted in considerable water savings.

However, not only co-benefits can occur, but also tradeoffs may need to be considered. Such tradeoffs can manifest in two ways: (1) Some technologies that can curb the power sector's carbon emissions may lead to increased water uses. For example, carbon capture and storage (CCS) offers the opportunity to minimize carbon emissions from fossil fuel power plants but at the expense of increased water uses for the parasitic loads (Zhai, Rubin and Versteeg, 2011; Byers et al., 2016). Although concentrated solar power (CSP) and inland nuclear power are both feasible alternatives to reduce carbon emissions of China's electric power sector, their on-site water uses are substantial, whereas bio-fuels require the largest amount of water inputs throughout the lifecycle, mainly for crop cultivation (Meldrum et al., 2013); (2) water-saving technologies at coal power plants can result in additional carbon emissions. For example, air-cooling technologies that use the least amount of water face energy penalties due to the lowered efficiency and thus increased level of greenhouse gas emissions (European Commission Joint Research Center, 2001).

Effects of cooling technology change

Responding to the emerging challenge, China has issued a series of policies targeted at managing sustainable water use at coal-fired power plants. At a national level, a new national water withdrawal standard (GB/T 18916.1–2012) was issued in 2012 to replace 'GB/T 18916.1–2002' and set more stringent requirements regarding water uses by coal power plants with closed-loop cooling systems (General Administration of Quality Supervision, Inspection and Quarantine of China, 2002, 2012). In water-scarce areas, new plants are not allowed to extract groundwater and are required to employ large air-cooling units (National Development and Reform Commission, 2004). As a result, new power plants are incentivized to adopt air-cooling systems so that their permits' application time can be shortened, and air-cooling units have proliferated consequently. The national percentage of air-cooling systems has increased from 6.4% in 2000 to 14.5% in 2015 (Figure 4.9). Such change demonstrated regional differences, with the closed-loop cooling system replacing the open-loop cooling system in the southern regions and air-cooling systems replacing wet cooling systems in the northern regions.

There are also other official but nonmandatory guidelines for water-use practices in the coal power sector, such as 'Guidelines for water saving in thermal power plants' issued by the State Economic and Trade Committee, 'China water conservation technology policy outline' by the NDRC and 'Water efficiency guide for key industrial sectors' by the Ministry of Industry and Information Technology (Zhang et al., 2016).

In total, changing cooling technologies has mitigated 14.07 billion m^3 of water withdrawal increase and 101.34 million m^3 of water consumption increase, which varied among regions. For instance, cooling technology change contributed to 93.94 million of water consumption increase but 5.10 billion m^3 of water withdrawal decrease in central China. While in northwest China, the percentage of closed-loop cooling systems has gone down from 86.6% in 2000 to 52.7% in 2015, being replaced by air-cooling systems, which resulted in 169.97 million m^3 of water consumption reductions. In the southern regions (central, south and east), closed-loop cooling systems are gradually replacing open-loop ones to reduce the power sector's water withdrawals and thus dependence on water supplies but at the cost of water consumption increase. While in the northern regions (north, northwest and northeast), in response to severe local water stresses,

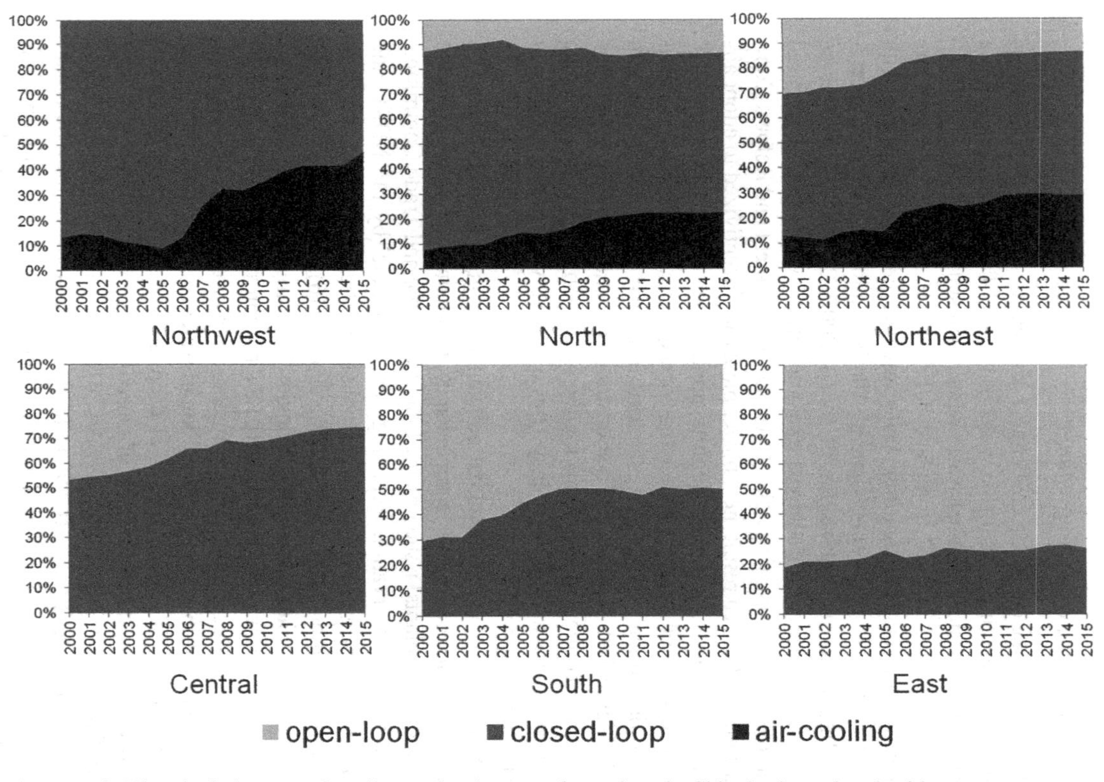

Figure 4.9 Historical changes of cooling technology configurations in China's six regional grids

Source: Liao and Hall, 2018.

air-cooling systems gradually have been adopted to cut both water withdrawal and consumption.

While replacing open-loop cooling systems with closed-loop ones reduces the power sector's dependency on water supplies and thus enhances its resilience to extreme weather events (e.g. droughts), it also markedly increases the power sector's consumptive water use and therefore competes with other water users (e.g. agriculture and domestic). Moreover, besides a larger capital investment (for land and cooling towers), more water-efficient technologies lead to lower power production efficiencies, which results in larger coal inputs and higher emissions, including carbon and other pollutants. Future infrastructure planning and investment need to consider those tradeoffs prudently.

Upgrading boiler technology may also contribute to the power sector's water savings. As Jiang and Ramaswami (2015) have demonstrated based on field data in China, EGUs with larger capacities are more water efficient. The reason is because larger EGUs tend to adopt more advanced boiler technologies (e.g. supercritical, with higher efficiencies) so that more heat is converted from the primary energy carrier (e.g. coal) to electricity, and therefore less residual heat needs to be dissipated by cooling water. Nonetheless their effects on water use are much smaller compared with plant types and cooling technologies (Macknick et al., 2012).

Future scenarios of water use by China's coal power plants

As China does not have any holistic plan for future electricity-generating technology development or infrastructure expansion (Li, 2015), Liao, Hall and Eyre (2016) used scenario analysis to manage the uncertainties in evaluating future contingencies. Four scenarios from the World Wide Fund for Nature (WWF) (2014) are examined where the baseline scenario (BS) indicates that China does not implement new policies other than the current ones; under the high-efficiency (HE) scenario, China reduces its energy intensity by transforming its economy from an energy-intensive industry to services and adopting state-of-the-art efficiency technologies. High-renewable (HR) and low-carbon (LC) scenarios require the demand to be met with renewable or low-carbon energy sources, respectively. As China's new 'Intended National Determined Contribution' (2015) pledged only to raise its share of non-fossil fuels in primary energy to about 20% in 2030, both the high-renewable and low-carbon scenarios are more optimistic than the current policy about China's changing fuel mix. Such a wide range of scenarios aim to demonstrate the impacts of different future energy portfolios on China's thermoelectric power sector's water uses instead of predicting the future. While in these

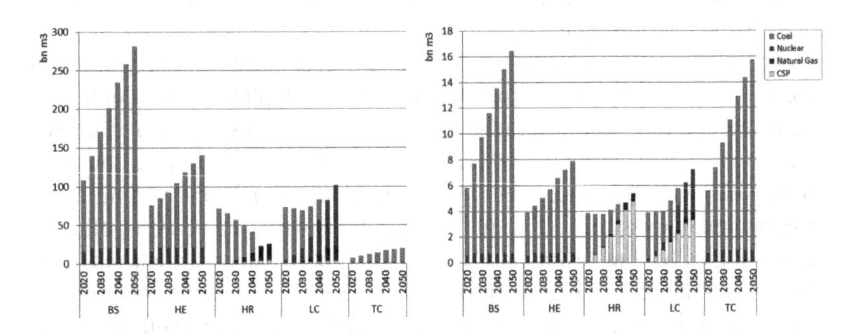

Figure 4.10 Freshwater withdrawal (left) and consumption (right) by China's thermoelectric power sector under different scenarios (billion m³)

Source: Liao, Hall and Eyre, 2016.

aforementioned scenarios, cooling technology configurations are assumed to remain the same over time, another additional scenario is developed to explore the effects of changing cooling technology (TC) configurations on the coal power sector's water use, where all closed-loop cooling systems in the north (north, northeast and northwest) are gradually replaced by air-cooling systems, and all open-loop cooling systems in the south (south, east and central) are changed to closed-loop ones.

Figure 4.10 shows that significant increases in water withdrawal and consumption can be seen in the baseline scenario to over 280 and 16 billion m³ respectively by 2050. Improving energy efficiency reduces the magnitudes of water withdrawal and consumption by over 50% to about 140 and 8 billion m³ by 2050 respectively.

Under high-renewable and low-carbon scenarios, water withdrawal is cut by over 70%, with a decreasing trend. Water consumption is also reduced by over 60% and grows at much lower rates. Under a low-carbon scenario, inland nuclear power is responsible for the growth in water consumption, requiring over 3.3 billion m³ by 2050.

Changing cooling technology significantly reduces water withdrawal due to the elimination of open-loop cooling systems. As for water consumption, phasing out open-loop cooling systems in the south and closed-loop cooling systems in the north, respectively, increases and decreases the sector's water consumption. However, little impact is seen on a national scale.

References

Bartos, M. D. and Chester, M. V. (2014). The conservation nexus: Valuing interdependent water and energy savings in Arizona. *Environmental Science & Technology*, 48, pp. 2139–2149.

Byers, E. A., Hall, J. W., Amezaga, J. M., ODonnell, G. M. and Leathard, A. (2016). Water and climate risks to power generation with carbon capture and storage. *Environmental Research Letters* 11: 024011.

The Center for Global Development. (2015). *Carbon monitoring for action*. Washington, DC: The Center for Global Development.

Chai, L., Liao, X., Yang, L. and Yan, X. (2018). Assessing life cycle water use and pollution of coal-fired power generation in China using input-output analysis. *Applied Energy* 231: 951–958.

China Electric Power Statistic Yearbook Editorial Board. (2015). *China electric power statistic yearbook 2000–2015* (in Chinese). Beijing, China: China Electric Power Press.

CoalSwarm. (2015). *Global coal plant tracker*. San Francisco: CoalSwarm.

Davis, C. B., Chmieliauskas, A., Dijkema, G. P. J. and Nikolic, I. (2015). *Enipedia*. Delft, The Netherlands: Energy & Industry Group, Faculty of Technology, Policy and Management, TU Delft. Available at: http://enipedia.tudelft.nl.

Diehl, T. H., Harris, M. A., Murphy, J. C., Hutson, S. S. and Ladd, D. E. (2013). *Methods for estimating water consumption for thermoelectric power plants in the United States scientific investigations report 2013–5188 US department of the interior*. Reston, VA: US Geological Survey.

Ehrlich, P. R. and Holdren, J. P. (1971). Impact of population growth. *Science*, 171, pp. 1212–1217.

European Commission Joint Research Centre. (2001). *Integrated pollution prevention and control (IPPC) reference document on the application of best available techniques to industrial cooling systems*. Seville, Spain: European Commission Joint Research Centre.

Falkenmark, M., Lundquist, J. and Widstrand, C. (1989). Macro-scale water scarcity requires micro-scale approaches: Aspects of vulnerability in semi-arid development. *Natural Resources Forum*, 13(4), pp. 258–267.

General Administration of Quality Supervision, Inspection and Quarantine of China, Standardization Administration of China. (2002). *Norm of water intake-part 1: Electric power production (GB/T 18916.1–2002)* (in Chinese). Beijing, China: General Administration of Quality Supervision, Inspection and Quarantine of China, Standardization Administration of China.

General Administration of Quality Supervision, Inspection and Quarantine of China, Standardization Administration of China. (2012). *Norm of water intake – part 1: Fossil fired power production (GB/T 18916.1–2012)* (in Chinese). Beijing, China: General Administration of Quality Supervision, Inspection and Quarantine of China, Standardization Administration of China.

Jiang, D. and Ramaswami, A. (2015). The 'thirsty' water-electricity nexus: Field data on the scale and seasonality of thermoelectric power generation's water intensity in China. *Environmental Research Letters*, 10, p. 024015.

Li, X. (2015). *Decarbonizing China's power system with wind power: The past and the future*. Oxford: The Oxford Institute for Energy Studies.

Liao, X., Hall, J. W. and Eyre, N. (2016). Water use in China's thermoelectric power sector. *Global Environmental Change*, 41, pp. 142–152.

Liao, X. W. and Hall, J. W. (2018). Drivers of water use in Chinas electric power sector from 2000 to 2015. *Environmental Research Letters* 13: 094010.

Macknick, J., Newmark, R., Heath, G. and Hallett, K. C. (2012). Operational water consumption and withdrawal factors for electricity generating technologies: A review of existing literature. *Environmental Research Letters*, 7, p. 045802.

Meldrum, J., Nettles-Anderson, S., Heath, G. and Macknick, J. (2013). Life cycle water use for electricity generation: A review and harmonization of literature estimates. *Environmental Research Letters*, 8, p. 015031.

Mi, Z. F., Meng, J., Guan, D. B., Shan, Y. L., Liu, Z., Wang, Y. T., Feng, K. S. and Wei, Y. M. (2017). Pattern change in determinants of Chinese emissions. *Environmental Research Letters*, 12, p. 074003.

National Development and Reform Commission of China (NDRC). (2004). *Requirements on the planning and construction of coal power plants* (in Chinese). Beijing, China: NDRC.

National Development and Reform Commission of China (NDRC). (2015). *Intended nationally determined contribution*. Beijing, China: NDRC.

National People's Congress of China. (2001). *The tenth five-year plan*. Beijing, China: National People's Congress of China.

National People's Congress of China. (2011). *The twelfth five-year plan*. Beijing, China: National People's Congress of China.

USA Today. (2013). *China endures worst heat wave in 140 years*. Available at: https://eu.usatoday.com/story/weather/2013/08/01/china-heat-wave/2608415/.

Utility Data Institute, Platts Energy InfoStore. (2015). *World electric power plants database*. Available at: www.platts.com.

Wang, X., Wu, S. and Li, S. (2017). Urban metabolism of three cities in Jing-Jin-Ji urban agglomeration, China: Using the MuSIASEM approach. *Sustainability*, 9(8), p. 1481.

Webster, M., Donohoo, P. and Palmintier, B. (2013). Water CO_2 trade-offs in electricity generation planning. *Nature Climate Change*, 3, pp. 1029–1032.

World Bank. (2017). *World Bank databank*. Washington, DC. Available at: https://data.worldbank.org/ indicator/EG.USE.ELEC.KH.PC.

World Wide Fund for Nature (WWF). (2014). *China's future power generation*. Beijing, China: WWF.

Zhai, H., Rubin, E. S. and Versteeg, P. L. (2011). Water use at pulverized coal power plants with post-combustion carbon capture and storage. *Environmental Science & Technology*, 45, pp. 2479–2485.

Zhang, C., Zhong, L., Fu, X. and Zhao, Z. (2016). Managing scarce water resources in China's coal power industry. *Environmental Management*, 57, pp. 1188–1203.

5 Water shortage risks for China's coal power plants

Current spatiotemporal distribution of water resources in China

China's total water resources rank sixth in the world, possessing 2,796 billion m^3 of annual renewable water resources in total. However, being home to nearly 20% of the global population (1.4 billion in 2018), its annual renewable water resources per capita are only 2008 m^3 (National Statistic Bureau, 2019), at only one-third of the global average, 5,930 m^3 per person (World Bank, 2019). Furthermore, China's water resources are spatially and temporally unevenly distributed. Rainfall is highest in the southeast and lowest in the northwest. The total average annual internal renewable surface water resources in northern areas are estimated to be 535.5 km^3, which is only 20% of the country's total (AQUASTAT, 2014). North China is home to 37% of its national population and 45% of the arable land but only 12% of the water resources (Figure 5.1).

The UN defines water resources per capita under 1,700 m^3as an indicator for water stress and under 1,000 and 500 m^3 for water scarcity and extreme water scarcity. Thirteen provinces, out of 31, in mainland China are facing water stresses, mainly in the north (Table 5.1), among which 12 are facing water scarcity and eight are facing extreme water scarcity. More than 650 million people now live in water-stressed areas in China, and the number is on the rise.

Such regionalized water scarcity is further aggravated by the mismatches with energy reserves. An estimated 71% of China's coal reserves are endowed in four water-scarce arid and semi-arid provinces of northern China, which in total account for only 13% of the national water resources.

To save the transportation costs of coal, coal power companies are incentivized to build coal power plants close to coal mines, which means over half of the country's coal-fired power capacity is located in water-stressed areas (China Water Risk, 2012; World Resource Institute, 2014) (Figure 5.2).

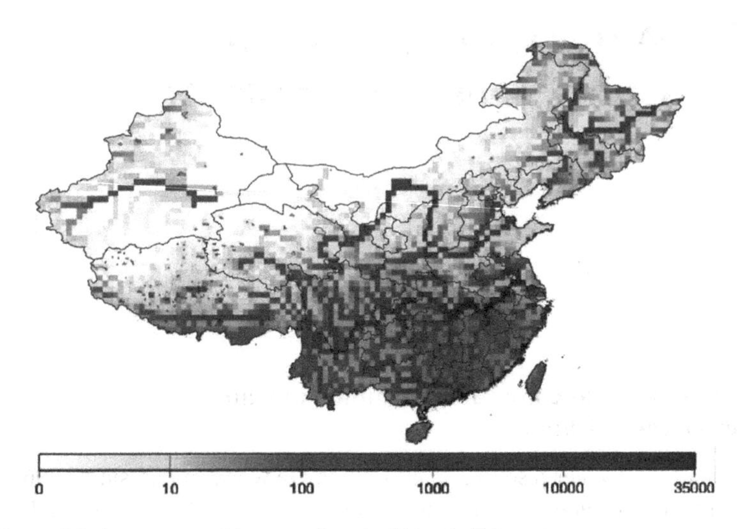

Figure 5.1 Average monthly mean flow in China (m^3/s)

Table 5.1 Provincial annual renewable water resources per capita in China

Region	Province	Water resources per capita (m^3/p)	Region	Province	Water resources per capita (m^3/p)
	Tianjin	121.58		Shanghai	252.33
	Ningxia	142.96		Jiangsu	928.58
	Beijing	161.60		Chongqing	1994.72
	Shandong	222.59		Anhui	2018.15
	Hebei	279.69		Guangdong	2250.64
	Henan	354.83		Zhejiang	2378.11
	Shanxi	365.10		Hubei	2552.61
North	Gansu	646.45	**South**	Sichuan	2843.31
	Shaanxi	713.91		Guizhou	3009.46
	Liaoning	757.08		Hunan	3229.11
	Inner Mongolia	1695.49		Yunnan	4391.67
	Jilin	1781.99		Guangxi	4522.73
	Heilongjiang	2217.05		Jiangxi	4850.62
	Xinjiang	4596.05		Hainan	5359.96
	Qinghai	10375.95		Fujian	5468.69
				Tibet	141746.56

Source: National Statistic Bureau, 2019.

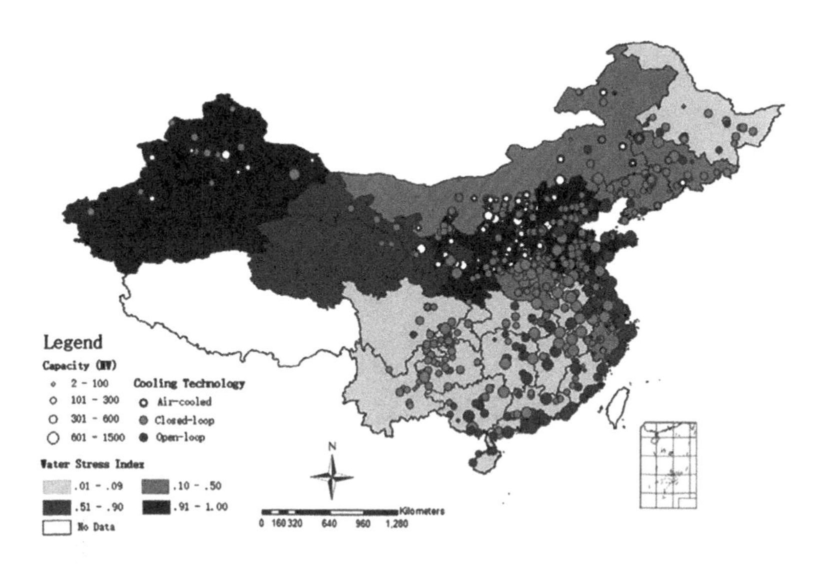

Figure 5.2 China's provincial water resource per capita and installed capacity (GW) from 2010 to 2015

Source: Chai et al., 2018.

Moreover, impacted by the monsoon climate, precipitation in China demonstrates substantial intra-annual variations, which is the lowest in the winter and highest in the summer (Ye and Zhang, 2013). Almost 70 to 80% of precipitation is concentrated from April to October, especially the two-to-three-month flood season during the summer, leaving the rest of the year relatively dry. Such intra-annual variations are greater in the north than in the south, therefore making winter in the north particularly vulnerable to drought hazards.

Seasonal pattern of China's electricity demands

The temperature in China varies considerably by region. In winter, the temperatures in northern areas (north, northwest and northeast) are below 0 °C and above zero in southern regions (central, south and east). In summer, most of the regions have temperatures above 20 °C. Such temperature differences directly result in regionally different electricity demand patterns and power plant cooling efficiencies. Hydropower production capacity is substantially higher in the summer due to larger amounts of stream flows.

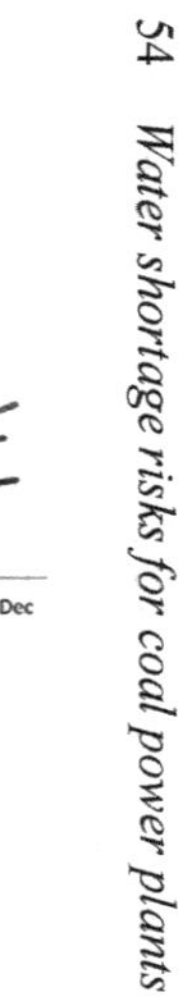

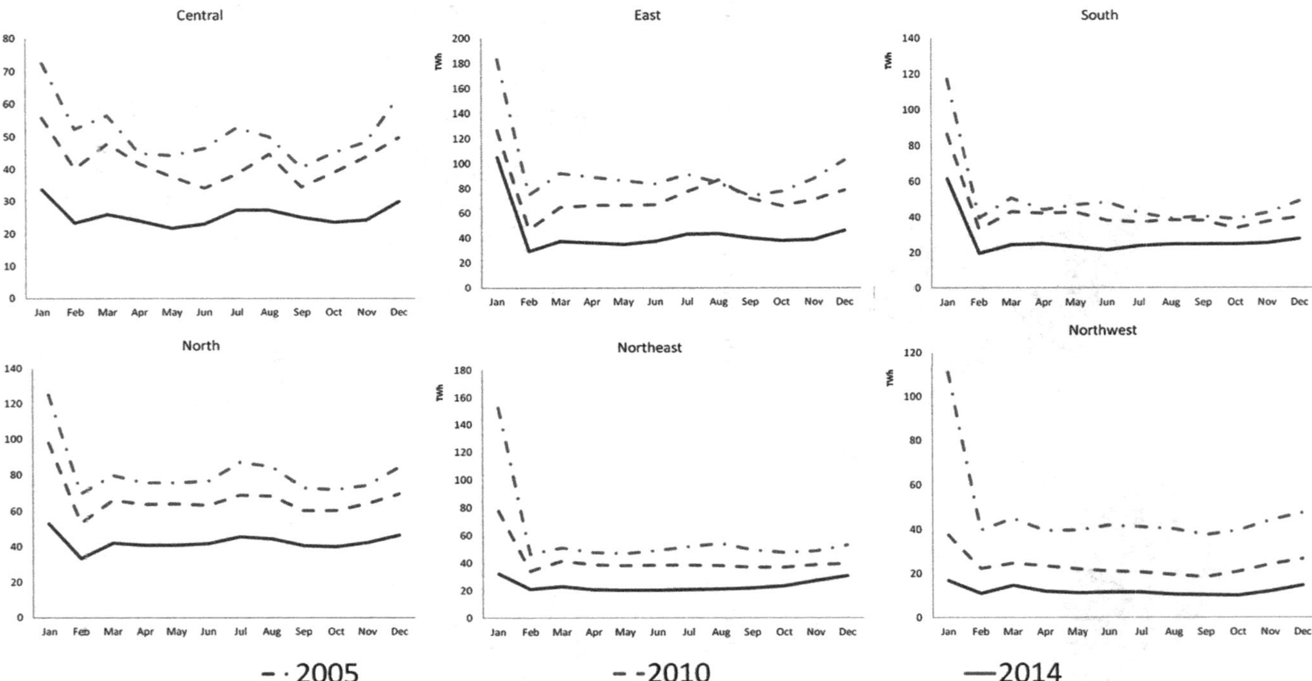

Figure 5.3 China's regional monthly thermal power productions from 2005 to 2014 (TWh)

Source: National Statistic Bureau, 2015.

China's regional monthly thermoelectricity production patterns can be seen from 2005 to 2014 (Figure 5.3). Apart from the overall growing trends, it can be seen that thermal power production peaks during the winter and summer throughout the whole of China. In winter, central heating by Combined Heat and Power is provided in northern areas from November to March, while heat pumps are prevalent in the other regions (i.e. central, east and south). In the summer, from June to August, electricity demand increases because of the widespread use of air conditioning and outweighs the increase in hydropower productions, especially in central and eastern China, and leads to thermoelectricity production peaks. The noticeable drops of power productions in February are presumably due to industrial shutdowns during the Chinese New Year.

Current water shortage risks for China's coal power plants

Coal power plants use water as a cooling medium to take away the residual heat in the exiting steam from steam turbines. The cooling efficiency can be affected by two factors: the volume of the water available and the temperature of the intake water. A smaller amount of water can realize the same cooling effect with a lower intake temperature, which is the primary reason why cooling water intensity is relatively small in winter compared to in summer (Jiang and Ramaswami, 2015). Often due to the restriction on discharge water temperature from coal power plants, if the intake water temperature is high, it requires a larger amount of cooling water to dissipate the same amount of residual heat. As a result, coal power production is exposed to potential water shortages through two changing mechanisms: water scarcity and increased intake water temperature. Incidents of thermal power production being disrupted by either intake or discharge water temperature being too high or insufficient intake water volume have been documented throughout the US (McCall, Macknick and Hillman, 2016). Sadoff et al. (2015) have pointed out that north China stands out as facing the highest water risks for the power productions in the world.

Focusing on the availability of intake water volume, Liao et al. (Under preparation) have compared the monthly water demand by coal power plants with available water resources in the river system during low flows and evaluated the current water shortage risks for China's coal power plants. Low flow is identified as being in the tenth percentile of river flows in a certain calendar month (monthly stream flow value that is exceeded 90% of the time in the same calendar month). At any given site of a power plant, if the low-flow water amount in the river is not enough to meet the water withdrawal demand by the power plant, the power plant is identified as facing low-flow water shortage risks. Their results show that about 20% of China's

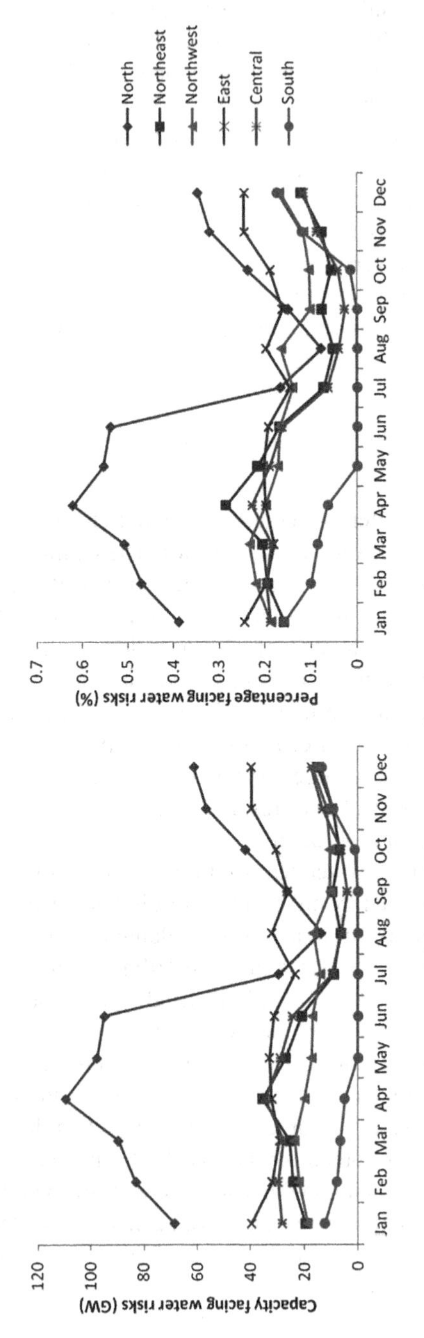

Figure 5.4 Monthly low-flow water risks for China's regional coal power plants

Source: Liao et al. under preparation.

national electricity generation capacity faces low-flow water shortage risks from November to June, with 30% facing low-flow water shortage risks in April. The exposed capacity is much lower from July to October, at only 10% on average (Figure 5.4). Such concentration from November to April results from two causes: First, water availability is lower during the winter and spring due to the monsoon repercussions; second, electricity demand is overall higher throughout China due to the requirements for heating purposes.

Looking at the regional differences, the largest proportion of thermoelectricity capacity faces water shortage risks in the north, while the least in the south. Over 60% of power generation capacity, whose aggregated capacity is larger than 100 GW, in the north is facing low-flow water risks if surface water is the only water source available (i.e. no groundwater utilization or reclaimed water use). It can be seen that such risks are concentrated in the Jing-Jin-Ji mega region (Figure 5.5) where the capital city Beijing is located.

In comparison, the risks for the southern grid's peak in December is mainly concentrated in the Yangtze Delta region when the heating demand is the highest throughout the year. The Jing-Jin-Ji and Yangtze Delta mega regions are China's two biggest and fastest-growing urban agglomerations. They are home to 111 and 158 million people, respectively, of which the urban population represents 62.5% and 69.8%. In 2016, these two regions generated over 10% and 20% of national GDP respectively.

Such water shortage risks are, respectively, supply-driven and demand-driven in the north and south. In the north grid, closed-loop cooling systems are most widely used with very small water withdrawal requirements.

Figure 5.5 Low-flow water shortage risks for China's coal power plants in April

Source: Liao et al., under preparation.

Therefore, the water shortage risks primarily resulted from the very low water availability from the river systems. As pointed out in Chapter 4, groundwater and reclaimed water are heavily used in the north China plain. According to China Electricity Council (2013a, 2013b), 16.2% and 12.2% of China's coal power plants' current water uses are groundwater and reclaimed water respectively. As the Chinese government has banned further groundwater utilization for newly built coal power plants to protect the over-exploited groundwater resources (NDRC, 2004), there are mainly two ways to lower such water shortage risks: (1) Increasingly deploy air-cooling units that require the least, almost negligible, amount of cooling water. Power plants switching to air-cooling systems can further reduce such water shortage risks but at the expense of higher energy penalties and emissions because of their lower cooling efficiency (Zhang et al., 2014). Air-cooling units also require larger capital investments and land; (2) the second solution is to increase the utilization of reclaimed water resources. China's newly issued 'National Water Saving Plan' set the target for the water recycling rate at large industrial plants to reach 91% by 2020 (NDRC and Ministry of Water Resources, 2019), which introduces additional costs for coal power plants.

In comparison, water risks in the south are primarily demand-driven. Although switching the widely used open-loop cooling systems to closed-loop cooling systems can be effective in reducing such risks during the winter, it might not be economically sensible, considering the increased costs and reduced efficiency, along with the large volumes of river flows during the summer. Further economic analysis should be carried out to evaluate such potential policy actions. Besides demand management, combining technologies with regulations and incentives could further help to manage demand.

Climate change's impacts on China's water resources

Global climate change is expected to increase rainfall variability, thus affecting water availability and variability for ecosystems and human use (O'Gorman and Schneider, 2009; Arnell, 2003). Many studies have shed light on the potential impacts in different river basins in China (Li et al., 2008; Yang et al., 2012). Leng et al. (2015) used a Variable Infiltration Capacity (VIC) model with climatic variables from the Fifth Assessment Report of the Intergovernmental Panel on Climate Change (IPCC, 2014) to simulate the potential changes of precipitation, evaporation and river runoffs throughout all of China. Their results show that due to the changes in precipitation and evapotranspiration, river flows are expected to grow in China's north from October to March, potentially alleviating the water shortage problems, whereas river flows are expected to decrease the rest of the year, particularly in May and June. Reverse trends can be expected in the south, with water availability decreasing during most of the year, except in September, when river flow is expected to increase marginally.

Liao et al. (under review) particularly examined China's four driest river basins, namely, Yellow, Huai, Hai and Inland River Basins, which are all located in the north and northwest. Hydrographs of the monthly river flows in these four basins under three time horizons (i.e. current, 2030 and 2050) are demonstrated in Figure 5.6, which shows the ensemble mean of five Global Circulation Models (GCMs) that are included in the Inter-Sectoral Impact Model Intercomparison Project (ISI-MIP) (Hempel et al., 2013).

It can be seen that, overall, wetter future conditions projected by the GCMs are expected to increase the mean water availability during low flows in the four river basins. The changing climate, at the projected time scale and for these analyzed scenarios, is not expected to alter the seasonal

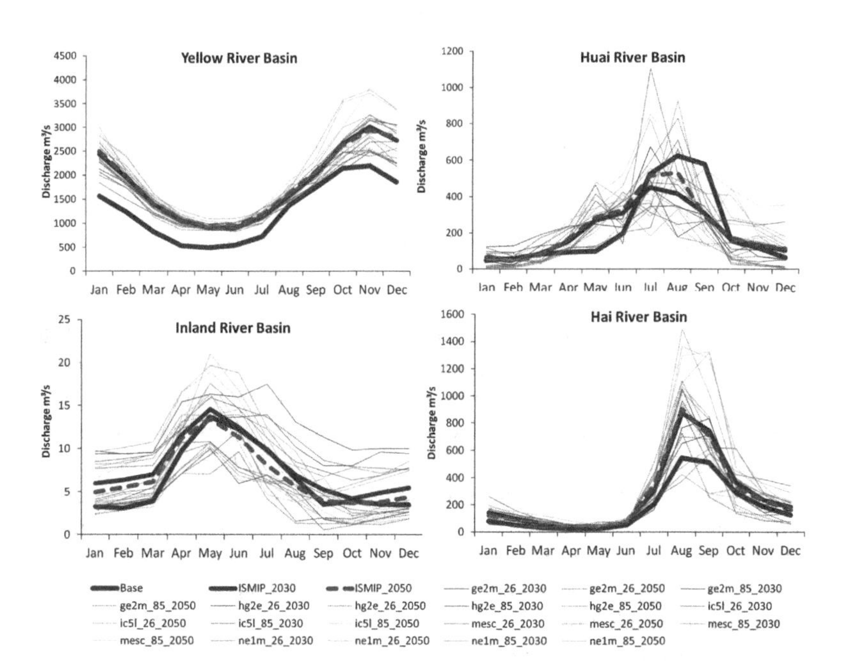

Figure 5.6 Monthly low-flow regimes in China's four driest river basins during the base period (1981–2014), future periods, 2030s (2015–2045) and 2050s (2035–2065). The results of five ISI-MIP GCMs are shown

Source: Liao et al., under preparation.

variability of river flows in these basins significantly. The driest months in the Huai and Inland River Basin are January and February, respectively, and May in both Hai and Yellow River Basin.

Future water shortage risks for China's coal power plants

By changing future water regimes, future climate change will also cause changes in water shortage risks faced by the coal power plants. Using a physically based hydrological model, a water temperature model and an electricity production model, Van Vliet et al. (2016) projected that over 80% of global thermal power plants' annual usable capacity can be reduced due to climate change during the period from 2040 to 2069. Moreover, climate change is also expected to affect electricity demand patterns and therefore the water demand for electricity productions. Zhou et al. (2018) estimated a similar trend as Van Vliet et al. (2016) by incorporating changes in water demand by thermal electric power production.

Liao et al. (under review) projected that, under the future wetter conditions, the water shortage risks facing China's coal power plants are expected to be attenuated in most regions (Figure 5.7). Only in the northwestern Inland River Basin is climate change expected to intensify low-flow water shortage risks for coal power plants from May to August unless the utilization rate of coal power facilities is lowered. Yet, an additional 5 to 10 GW of coal power facilities within this region will risk not having enough water for cooling during the summer months if their utilization rate is kept the same or increased.

Although water shortage risks for coal power plants are generally small in the east and south China, if the utilization rates of their coal power plants are increased, for instance, under circumstances where their increasing regional electricity demands remain highly dependent on coal, their coal power plants are expected to face heightened risks of low river flows not being sufficient to meet their withdrawal demands, in February in the east and in March and November in the south. Power plants are threatened to curtail their production, with consequential impacts for electricity revenues. Retrofitting open-loop cooling systems to closed-loop ones in the south and east is an effective investment to reduce their corresponding water risks and potential economic loss.

In order to reduce dependence on coal power plants, the electric power system needs to transition to other energy sources, such as wind, solar and hydropower. Although China's renewable energy has grown at a remarkable rate during the last decade, with costs for solar power and wind power

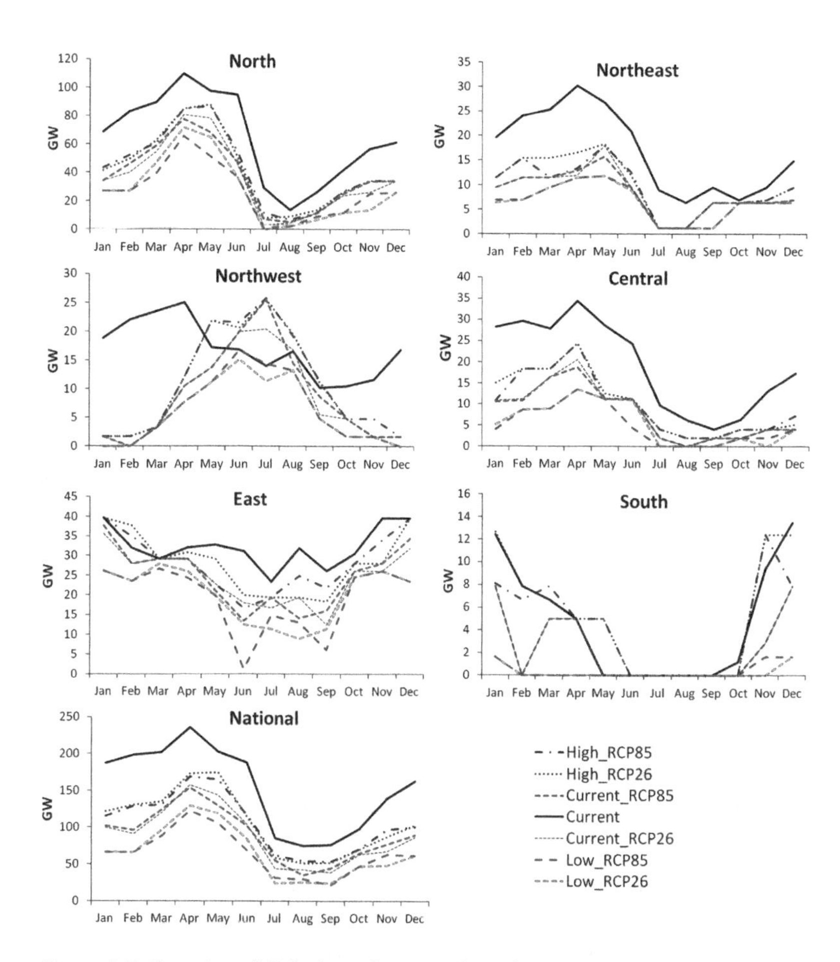

Figure 5.7 Capacity of China's coal power plants facing low-flow water risks during the base (2000s) and future (2050s) periods

Source: Liao et al., under preparation.

reducing to the grid parity, there are still challenges associated with intermittency and geographical mismatch between inland provinces where wind, hydro and solar resources are most abundant and coastal provinces where energy demand is greatest. The solution lies in institutional change to facilitate electricity transmissions from those electricity-producing regions to the load centers. The next chapter will shed light on the effects of inter-regional electricity transmissions on water uses for coal power plants.

References

AQUASTAT. (2014). *AQUASTAT country profiles*. Rome, Italy: FAO.

Arnell, N. W. (2003). Effects of IPCC SRES emissions scenarios on river runoff: A global perspective. *Hydrology and Earth System Sciences*, 7, pp. 619–641.

Chai, L., Liao, X., Yang, L. and Yan, X. (2018). Assessing life cycle water use and pollution of coal-fired power generation in China using input-output analysis. *Applied Energy*, 231, pp. 951–958.

China Electricity Council (CEC). (2013a). *National 600 Mw scale thermal power unit benchmarking and competition dataset* (in Chinese). Beijing, China: China Electricity Council.

China Electricity Council (CEC). (2013b). *National 300 Mw scale thermal power unit benchmarking and competition dataset* (in Chinese). Beijing, China: China Electricity Council.

China Water Risk. (2012). *China: No water, no power*. Available at: http://chinawaterrisk.org/resources/analysis-reviews/china-no-water-no-power/.

Hempel, S., Frieler, K., Warszawski, L., Schewe, J. and Piontek, F. (2013). A trend-preserving bias correction – the ISI-MIP approach. *Earth System Dynamics*, 4, pp. 219–236.

IPCC. (2014). *Climate change 2014: Synthesis report*. Core Writing Team, R. K. Pachauri and L. A. Meyer, eds. Contribution of Working Groups I, II and III to the Fifth Assessment Report of the Intergovernmental Panel on Climate Change. Geneva, Switzerland: IPCC, p. 151.

Jiang, D. and Ramaswami, A. (2015). The 'thirsty' water-electricity nexus: Field data on the scale and seasonality of thermoelectric power generation's water intensity in China. *Environmental Research Letters*, 10, p. 024015.

Leng, G. Y., Tang, Q. H., Huang, M. Y., Hong, Y. and Ruby, L. L. (2015). Projected changes in mean and interannual variability of surface water over continental China. *Science China Earth Sciences*, 58(5), pp. 739–754.

Li, L., Hao, Z. C., Wang, J. H. and Yu, Z. B. (2008). Impact of future climate change on runoff in the head region of the Yellow River. *Journal of Hydrologic Engineering*, 13, pp. 347–354.

Liao, X., Hall, J. W., Naota, H., Lim, W. and Paltan, H. (under preparation). *Water shortage risks in China's coal power industry under climate change*.

McCall, J., Macknick, J. and Hillman, D. (2016). *Water-related power plants curtailments: An overview of incidents and contributing factors*. Golden, CO: National Renewable Energy Laboratory.

National Development and Reform Commission of China. (2004). *Requirements on the planning and construction of coal power plants* (in Chinese). Available at: www.nea.gov.cn/2012-01/04/c_131262602.html.

National Development and Reform Commission of China and Ministry of Water Resources. (2019). *National water saving plan*. Beijing, China: NDRC, MWR.

National Statistic Bureau. (2015). *Statistic yearbook of China*. Beijing, China: National Statistic Bureau.

National Statistic Bureau. (2019). *Statistic yearbook of China*. Beijing, China: National Statistic Bureau.

O'Gorman, P. A. O. and Schneider, T. (2009). The physical basis for increases in precipitation extremes in simulations of 21st-century climate change. *Proceedings of the National Academy of Sciences of the United States of America*, 106, pp. 14773–14777.

Sadoff, C. W., Hall, J. W., Grey, D., Aerts, J. C. J. H., Ait-Kadi, M., Brown, C., Cox, A., Dadson, S., Garrick, D., Kelman, J., McCornick, P., Ringler, C., Rosegrant, M., Whittington, D. and Wiberg, D. (2015). *Securing water, sustaining growth: Report of the GWP/OECD task force on water security and sustainable growth.* Oxford: University of Oxford Press.

Van Vliet, M. T. H., Wiberg, D., Leduc, S. and Riahi, K. (2016). Power-generation system vulnerability and adaptation to changes in climate and water resources. *Nature Climate Change*, 6, pp. 375–380.

World Bank. (2019). *World Bank databank.* Washington, DC. Available at: https://data.worldbank.org/indicator.

World Resource Institute (WRI). (2014). *Identifying the global coal industry's water risks*. Available at: www.wri.org/blog/2014/04/identifying-global-coal-industry%E2%80%99s-water-risks.

Yang, C., Yu, Z., Hao, Z., Zhang, J. and Zhu, J. (2012). Impact of climate change on flood and drought events in Huaihe River Basin, China. *Hydrology Research*, 43, pp. 14–22.

Ye, W. H. and Zhang, Y. (2013). *Environment management.* 3rd Version (in Chinese). Beijing, China: China Higher Education Press.

Zhang, C., Anadon, L. D., Mo, H., Zhao, Z. and Liu, Z. (2014). Water-carbon trade-off in China's coal power industry. *Environmental Science & Technology*, 48(19), pp. 11082–11089.

Zhou, Q., Hanasaki, N., Fujimori, S., Yoshikawa, S., Kanae, S. and Okadera, T. (2018). Cooling water sufficiency in a warming world: Projection using an integrated assessment model and a global hydrological model. *Water*, 10(7), p. 872.

6 Water impacts of electricity transmissions in China

Mismatched electricity consumption and production in China

Besides being the world's largest electricity producer and consumer, electricity consumption and production are spatially imbalanced in China. Its electricity consumption is concentrated along the densely populated and economically advanced coastal regions, especially in the three major megalopolises (Figure 6.1): (1) Jing-Jin-Ji Bohai Economic Rim in the north where the capital city Beijing is located; (2) the Yangtze River Delta in the east where China's financial center Shanghai is located; and (3) the Pearl River Delta in the south, bordering Hong Kong, where China's first Special Economic Zone Shenzhen is located. The formation of megalopolises, which are defined as regions of adjacent heavily populated metropolitan cities, facilitates and symbolizes the processes of urbanization (Briggs, 2015). In such megalopolises, an increasingly concentrated population, elevated levels of economic activities and the burgeoning lifestyle of urbanites all require higher levels of electricity consumption. As shown in Figure 6.1, those three megalopolises all have populations larger than 100 million, of which over 60% are urbanites, and together they make up over 40% of the national GDP.

On the other hand, electricity production extracts energy in primary energy carriers (e.g. coal and hydropower resources), which are mostly endowed far from the coastal areas. To solve such spatial mismatches, China has built some large-scale inter-regional electricity transmission infrastructures since the late 1970s. For example, the 'West-to-East' electricity transmission project (hereinafter referred to as WETP project) was initiated in China's 10th Five-year Plan in 2001 (National People's Congress, 2001) and designed to bring economic development to the lagging west while alleviating the resources pressure in the east. The WETP project consists of three corridors: (1) The south corridor transfers hydropower generated from

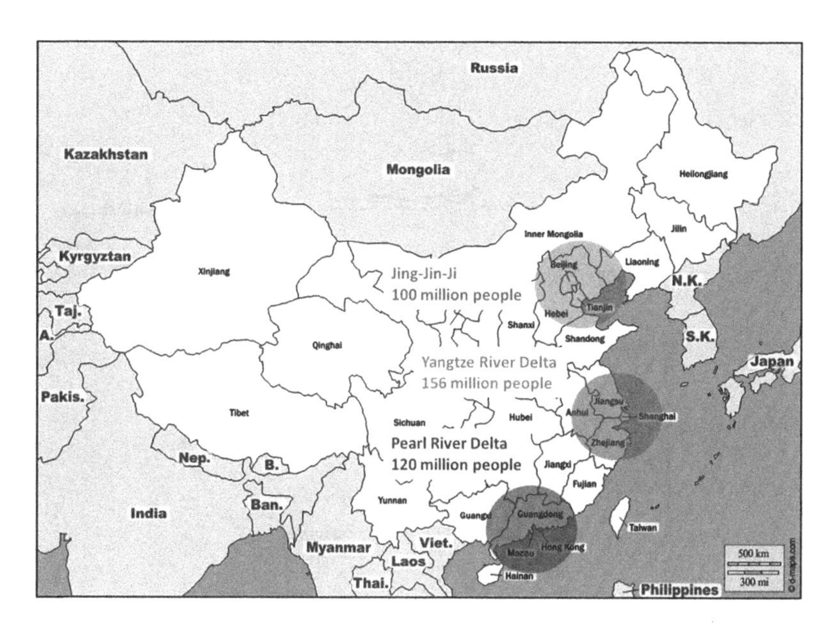

Figure 6.1 Three megalopolises in China

southwest regions to the Pearl River Delta; (2) the central corridor transfers hydropower from central regions and coal power from the northwest to the Yangtze River Delta; (3) the north corridor transfers electricity from the coal-abundant northwest to the Jing-Jin-Ji Bohai Economic Rim. As of November 2018, the State Grid has built 21 ultra-high voltage (UHV) transmission lines, and the inter-provincial transmission capacity has exceeded 140 million kilowatts (National Energy Administration, 2018). The disparities between regional electricity production and consumption are evident in Figure 6.2, indicating inter-regional electricity transmissions.

As has been seen in previous chapters, water is an important input to coal power production, and water scarcity poses increasing constraints and risks for coal power plants, particularly in China's arid and semi-arid northern regions. However, China's coal reserves are largely endowed in those northern regions. For instance, more than 50% of the coal reserves in China are endowed in the Yellow River Basin where water scarcity plays an essential constraint in social-economic development. Both the central corridor and north corridor of the WETP project heavily utilize electricity generated from coal power plants in the northern regions to support the consumption on the coast, which is further strengthened by the 12 new long-distance electricity

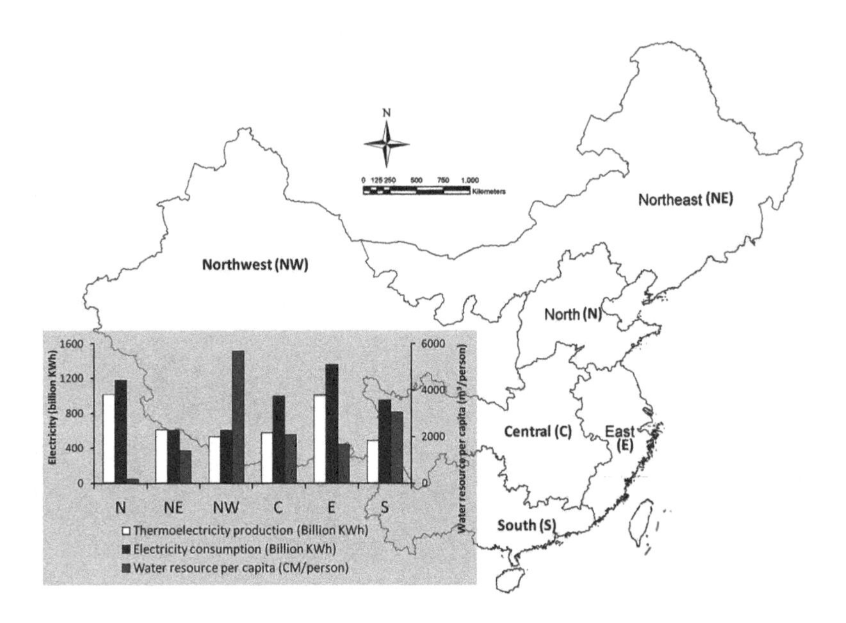

Figure 6.2 Regional electricity production (TWh), electricity consumption (TWh) and available water resource per capita (m^3/year) in China's regional power grids

Source: National Statistic Bureau, 2018.

transmission lines initiated in 2014 by China's National Energy Administration, whose aggregated total capacity amounts to 72 GW (National Energy Administration, 2014).

As can be seen from Figure 6.3, apart from the south corridor that utilizes hydropower from water-abundant southwestern China, the other 11 transmission lines all transfer electricity produced from rich coal resources in provinces that suffer from water stresses (i.e. Shanxi, Shaanxi, Ningxia and Inner Mongolia). Since electricity generation in the electricity-exporting provinces use local water resources, electricity transmissions shift such negative ecological and environmental impacts from electricity-importing provinces to exporting ones. Overall, electricity transmission generates impacts on water resources via three mechanisms: (1) It imposes pressure on water stress in the electricity-exporting provinces because water is used to produce electricity; (2) it reduces water consumption in the electricity-importing provinces compared to the counterfactual scenario where those

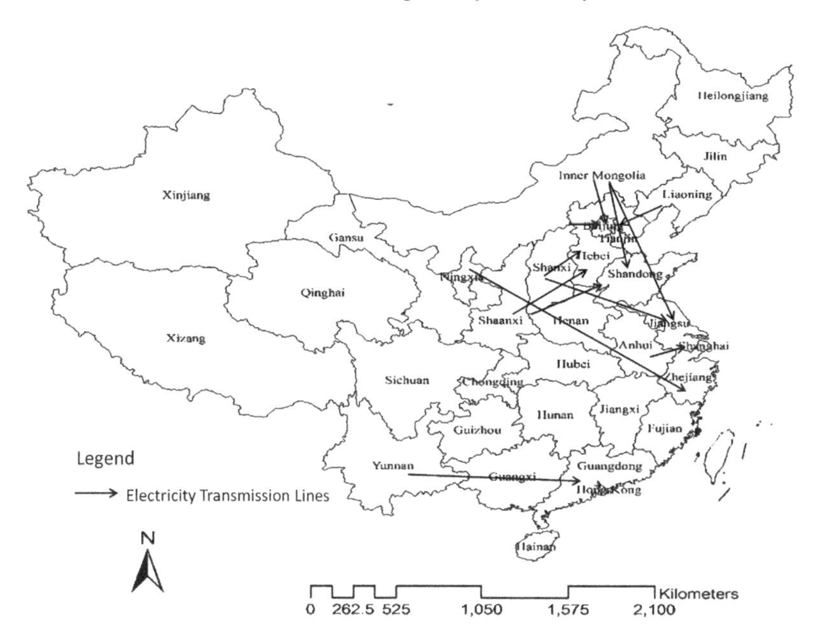

Figure 6.3 The 12 proposed electricity transmission lines in China and local water stresses

provinces would need to produce the electricity by themselves; (3) on a national level, if the exporting region has higher water productivity than the importing region, water savings can be generated on an aggregated national level; otherwise, additional water resources are consumed.

Water impacts of electricity transmission in China in 2014

According to Liao et al. (2019), 6.31 and 0.622 km³ of water was withdrawn and consumed, respectively, to produce the transmitted coal-fired power in all electricity-exporting provinces in China in 2014. It should be noted that water consumption for hydropower production is not included in their study. The amounts of electricity transmission from coal power plants were extrapolated based on each province's total power transmission and their energy mix data. As can be seen from Figure 6.4, most water was withdrawn in the central (2.11 km³) and east (2.12 km³), primarily due to the prevalence of open-loop cooling systems, while the largest amount of water was consumed in northeastern provinces at a total of 0.21 km³.

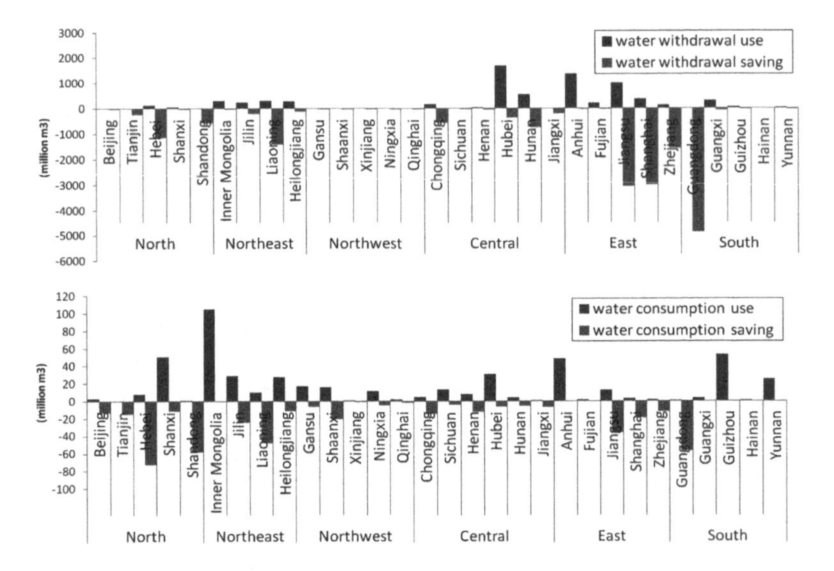

Figure 6.4 Provincial water uses and savings by China's inter-provincial electricity transmissions in 2014

In 2014, 26.41 km³ of water withdrawal and 0.64 km³ of water consumption were avoided in electricity-receiving provinces by importing electricity. On a national level, due to the differences in water productivities, 20.10 km³ of water withdrawal savings and 0.021 km³ of water consumption savings were realized by inter-provincial electricity transmissions in China.

Water impacts of the 12 proposed electricity transmission lines

According to Peng et al. (2017), the detailed information of the 12 newly proposed long-distance power transmission lines is presented in Table 6.1.

The water impacts of the 12 proposed electricity transmission lines are analyzed under two utilization scenarios of these transmission lines, at an 80% (high) utilization rate and a 50% (low) utilization rate. Since there are two lines designed to transmit both wind and coal-fired power (i.e. lines 4 and 5), three scenarios are considered regarding the energy mix for producing the transmitted electricity in the exporting provinces: (1) coal max (CM), where all transmitted electricity is produced by coal; (2) wind max

Table 6.1 Detailed information of the 12 long-distance electricity transmission lines

#	*Exporting provinces*	*Importing provinces*	*Capacity (GW)*	*Energy source*	*Completion year*
1	Liaoning	Beijing, Tianjin, Hebei & Shandong	2	Coal	2015
2	Shaanxi	Hebei	3	Coal	2016
3	Shanxi	Hebei	3	Coal	2017
4	Inner Mongolia	Beijing, Tianjin, Hebei & Shandong	9	Coal + Wind	2017
5	Inner Mongolia & Shanxi	Beijing, Tianjin & Hebei	6	Coal + Wind	2017
6	Shaanxi & Shanxi	Hebei & Shandong	6	Coal	2017
7	Inner Mongolia	Shandong	8	Coal	2016
8	Anhui	Jiangsu, Zhejiang & Shanghai	8	Coal	2016
9	Ningxia	Zhejiang	8	Coal	2016
10	Inner Mongolia	Jiangsu	7	Coal	2017
11	Shanxi	Jiangsu	7	Coal	2017
12	Yunnan	Guangdong	5	Hydropower	2017

(WM), where wind power in the two exporting provinces is used to its maximum capacity, and (3) half-half (HH), where equal amounts of wind power and coal-fired power are transmitted.

Table 6.2 shows that when the long-distance transmission lines are operated at a utilization rate of 80%, from 391.3 (WM) to 497.6 (CM) million m^3 of water consumption and from 2.97 (WM) to 3.24 (CM) billion m^3 of water withdrawal will be induced for electricity transmissions by the 12 transmission lines in total depending on how much wind power will be used in the Inner Mongolia and Shanxi provinces. The largest amount of water will be withdrawn in Anhui, 1.99 billion m^3.

It can be seen from Figure 6.3 that although some of the electricity-exporting regions are facing severe water scarcities, almost all of the electricity-importing regions are facing even more severe water scarcities. By importing transmitted electricity, water uses are avoided in the importing regions that otherwise would need to be used. The potential water savings by the 12 electricity transmission lines are illustrated in Figure 6.5. Regarding water consumption, electricity transmission lines to provinces in the north all incur net water savings, while electricity transmission lines to provinces in the Yangtze Delta region increase net water consumption on a national level. The reason is because, while power production in the exporting regions mostly employs closed-loop cooling systems that

Table 6.2 Water uses (million m³) to produce transmitted electricity in the exporting provinces at a utilization rate of 80%

Long-distance power transmission lines		*Water withdrawal*			*Water consumption*		
Exporting provinces	*Importing provinces*	*CM*	*WM*	*HH*	*CM*	*WM*	*HH*
Liaoning	Beijing, Tianjin, Hebei & Shandong	461.70	461.70	461.70	17.28	17.28	17.28
Shaanxi	Hebei	25.46	25.46	25.46	21.32	21.32	21.32
Shanxi	Hebei	21.21	21.21	21.21	18.01	18.01	18.01
Inner Mongolia	Beijing, Tianjin, Hebei & Shandong	174.92	0.00	87.46	65.67	0.00	32.83
Inner Mongolia & Shanxi	Beijing, Tianjin & Hebei	86.60	0.00	43.30	40.64	0.00	20.32
Shaanxi & Shanxi	Hebei & Shandong	45.84	45.84	45.84	38.68	38.68	38.68
Inner Mongolia	Shandong	155.48	155.48	155.48	58.37	58.37	58.37
Anhui	Jiangsu, Zhejiang & Shanghai	1990.88	1990.88	1990.88	71.73	71.73	71.73
Ningxia	Zhejiang	88.39	88.39	88.39	72.82	72.82	72.82
Inner Mongolia	Jiangsu	136.05	136.05	136.05	51.08	51.08	51.08
Shanxi	Jiangsu	49.48	49.48	49.48	42.01	42.01	42.01

require large water consumption, power plants in the Yangtze Delta region prevalently use open-loop cooling systems; therefore, water consumption avoided is minimal. On a national total, from 58.21 (CM) to 164.52 (WM) million m³ of water consumption will be saved depending on different wind use rates.

In terms of water withdrawal, on a national aggregated level, assuming all lines' utilization rate is 80%, 17.90 (CM) to 18.16 (WM) billion m³ of water withdrawal will be saved by the 12 proposed long-distance transmission lines.

Accounting for both volumetric water and scarce water

The impacts of water use depend not only on the volume of water withdrawn but also on the availability of water where the water is used. Water

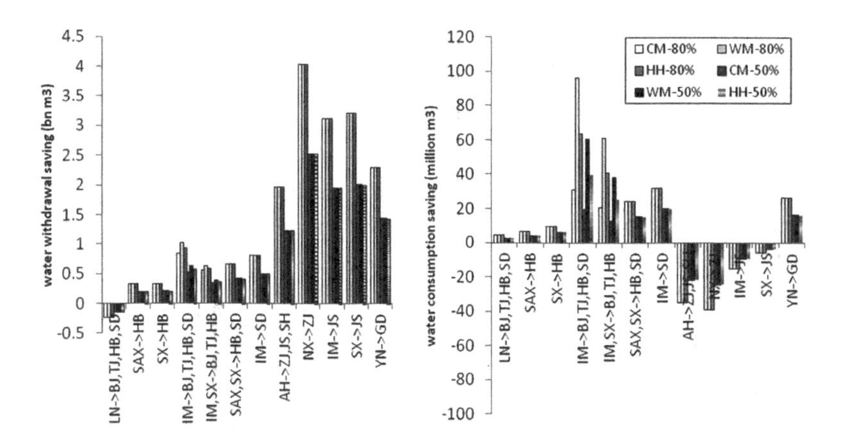

Figure 6.5 Net water use/savings by proposed long-distance electricity transmission lines

Note: LN – Liaoning; BJ – Beijing; TJ – Tianjin; HB – Hebei; SD – Shandong; SAX – Shaanxi; SX – Shanxi; IM – Inner Mongolia; AH – Anhui; JS – Jiangsu; ZJ – Zhejiang; SH – Shanghai; NX – Ningxia; YN – Yunnan; GD – Guangdong

used in the water-scarce north China has much larger impacts than the same amount of water used in the water-abundant south, on both the environment (e.g. river ecological system) and other water users in human society. To account for local water scarcities, both scarce water uses and scarce water savings are estimated by multiplying the Water Stress Index (WSI) developed by Pfister, Koehler and Hellweg (2009), which is multi-year average values derived from a global gridded hydrological model. WSI adjusts a region's water scarcity into a dimensionless continuous value between 0.01 and 1. A higher value of WSI indicates higher water stress. A WSI of 0.5 is usually regarded as a threshold between medium and higher water stress (Gassert et al., 2014).

In 2014, 1.46 km^3 of scarce water was withdrawn in the electricity-exporting provinces to produce the electricity that is transmitted, among which 0.59 km^3 was withdrawn from the north in spite of its dry conditions. Although water is relatively abundant in the south, 0.86 km^3 of scarce water was still withdrawn to produce electricity for exporting purposes, which highlights the importance of improving water productivity even in water-abundant regions. However, 12.44 km^3 of scarce water was saved in electricity-receiving provinces by importing electricity. On a national level, 10.98 km^3 of scarce water was saved due to all inter-provincial electricity transmissions.

The magnitudes of scarce water use and savings achievable from proposed electricity transmission lines are much smaller. When the transmission lines operate at a higher utilization rate (80%) and both coal and wind are equally utilized for lines 4 and 5, 157.17 and 610.73 million m^3 of scarce water will be consumed and withdrawn, respectively, in the electricity-exporting provinces. After being either consumed or withdrawn for electricity production, physical water can be thought of as being transformed into virtual water embodied in the transmitted electricity, which can then be transferred across administrative boundaries. Virtual water was introduced by Allan (1993) to measure the water used in the production of goods and services that are then traded to other places. Importing goods and services containing virtual water helps to alleviate water use in the place where they are finally consumed, but creates an added burden of water withdrawal in the places where the traded goods and services are produced. The virtual scarce water consumption transfers with the 12 proposed long-distance electricity transmission lines are illustrated as an example in Figure 6.6.

Three out of the six exporting regions, Ningxia, Liaoning and Shanxi, are facing severe water scarcity (WSI > 0.5). Especially in Ningxia, its water use has exceeded its annual renewable freshwater resources, indicating a heavy utilization of alternative water resources (e.g. groundwater). Nonetheless, the largest amount of scarce water consumption, 72.82 million m^3, will be transferred outward from Ningxia. On the other hand, Liaoning will be exporting the largest amount of scarce water withdrawal of 280.73 million m^3. In total, when both coal and wind power are used from Inner Mongolia and Shanxi, and all lines' utilization rate is 80%, 11.51 billion m^3 of scarce water withdrawal and

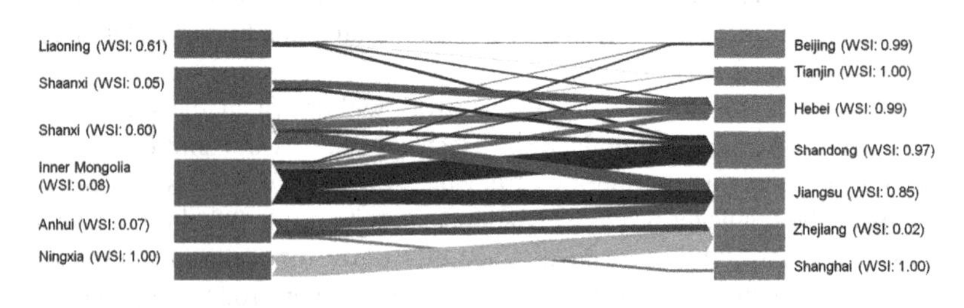

Figure 6.6 Fluxes of virtual scarce water consumption transfers with the 12 proposed electricity transmission lines

Note: Flux length is proportional to the water amount.

306.83 million m³ of scarce water consumption will be saved by the 12 proposed transmission lines.

Electricity transmissions' impacts on local water resources

Electricity transmissions' impacts on local water resources are analyzed by adding the water impacts of electricity transmissions in 2014 to those of the 12 proposed transmission lines, assuming their utilization rate is at 80% and under the HH scenario. It can be seen from Table 6.3 that, in Ningxia, Liaoning and Shanxi, where there are severe water stresses, water use for electricity exports accounts for more than 1% of its total provincial water use. Moreover, although Anhui does not have serious water stresses (WSI=0.07), water use for electricity exports makes up more than 10% of its provincial total water use.

Among all the power-importing provinces that face serious water stresses, counterfactual water savings due to electricity imports represent a substantial proportion of provincial water uses. Taking Shanghai, for example, which is already facing extreme water scarcity (WSI=1), where, were it to produce all the electricity it imports, the corresponding water use would equal over 40% of its current water use.

Drivers of water consumption for electricity transmission

By applying a Quasi-Input-Output model (Qu et al., 2017) coupled with a Structural Decomposition Model (Su and Ang, 2012; Gao et al., 2019), Liu et al. (2020) identified and quantified the impacts of five drivers of water uses for electricity transmissions in China, namely (1) transmission amount per GDP; (2) GDP; (3) transmission structure; (4) power generation structure and (5) water use intensities for electricity generation.

The impacts of those drivers for water consumption change are demonstrated in Figure 6.7 as an example. It can be seen that GDP growth has played the primary driver for the increase in water consumption for electricity transmissions in China, but its effect has been decreasing. From 2014 to 2016, the amount of electricity transmission has increased at a higher rate than GDP and has generated the largest impact on the associated water consumption increases. Changes in power generation structure and water intensity have played offsetting roles due to the transition away from coal power plants and uptake of water-saving technology (e.g. air-cooling systems.

Table 6.3 Provincial water uses/savings* for electricity transmissions compared to provincial total water use

Exporting provinces	*Water uses for exporting electricity (km³)*	*Total provincial water use** (km³)*	*Percentage in provincial water use*	*WSI*	*Importing provinces*	*Water savings by importing electricity (km³)*	*Total provincial water use (km³)*	*Percentage in provincial water use*	*WSI*
Ningxia	0.13	7.12	1.9%	1.00	Tianjin	0.40	2.40	16.6%	1.00
Liaoning	0.46	14.23	3.2%	0.61	Shanghai	4.97	11.47	43.3%	1.00
Shanxi	0.19	7.33	2.5%	0.60	Beijing	0.07	3.68	2.0%	0.99
Gansu	0.02	12.15	0.2%	0.51	Hebei	3.73	19.25	19.4%	0.99
Xinjiang	0.02	57.21	0.0%	0.35	Shandong	2.65	21.82	12.1%	0.97
Heilongjiang	0.09	35.86	0.3%	0.14	Jiangsu	12.31	57.02	21.6%	0.85
Jilin	0.11	13.18	0.8%	0.12	Liaoning	1.07	14.23	7.5%	0.61
Inner Mongolia	0.89	18.40	4.8%	0.08	Henan	0.07	22.81	0.3%	0.59
Anhui	3.34	28.88	11.6%	0.07	Shaanxi	0.00	8.92	0.0%	0.05
Hubei	1.20	29.55	4.0%	0.06	Hainan	0.04	4.48	0.9%	0.04
Shaanxi	0.05	8.92	0.5%	0.05	Guangdong	9.32	44.88	20.8%	0.04
Fujian	0.06	20.41	0.3%	0.03	Chongqing	0.31	8.26	3.7%	0.03
Guangxi	0.12	30.40	0.4%	0.02	Hunan	0.61	33.01	1.8%	0.03
Sichuan	0.05	24.48	0.2%	0.02	Zhejiang	8.93	19.48	45.9%	0.02
Guizhou	0.09	9.63	1.0%	0.02	Jiangxi	0.34	25.51	1.3%	0.02
Yunnan	0.03	14.96	0.2%	0.02	Qinghai	0.01	2.80	0.2%	0.01

* Water uses and savings in this table are the sum of the current amount and future amount for the 12 proposed transmission lines

** Provincial total water use is a five-year average calculated based on data from 2011 to 2015

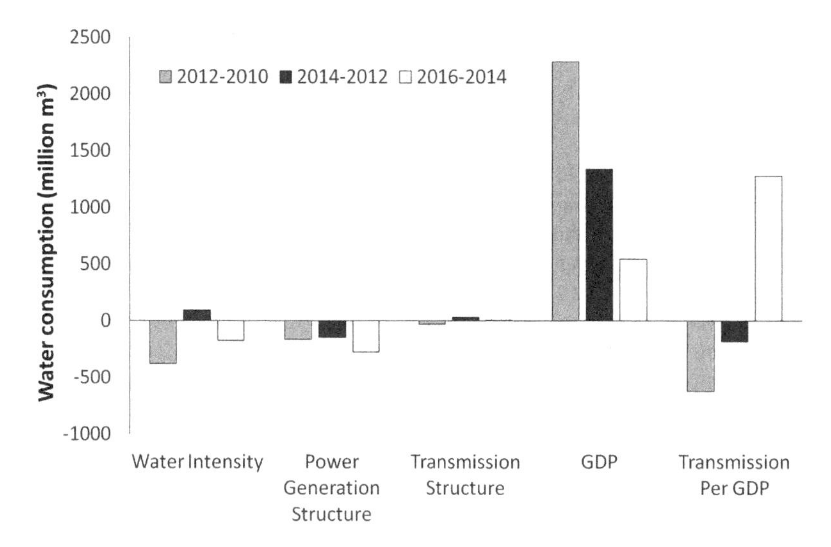

Figure 6.7 Drivers of water consumption for electricity transmissions in China

References

Allan, J. A. (1993). Fortunately there are substitutes for water otherwise our hydro-political futures would be impossible In: *Priorities for water resources allocation and management*. London: ODA, pp. 13–26.

Briggs, J. (2015). Capitals, Wizards owner Ted Leonsis says Baltimore will become part of a D.C. supercity. *Baltimore Business Journal News*, Baltimore.

Gao, H., Gu, A., Wang, G. and Teng, F. (2019). A structural decomposition analysis of China's consumption-based greenhouse gas emissions. *Energies*, 12, p. 2843.

Gassert, F., Luck, M., Landis, M., Reig, P. and Shiao, T. (2014). *Aqueduct global maps 2.1: Constructing decision-relevant global water risk indicators*. Washington, DC: World Resources Institute.

Liao, X. W., Chai, L., Jiang, Y., Ji, J. P. and Zhao, X. (2019). Inter-provincial electricity transmissions' co-benefit of national water savings in China. *Journal of Cleaner Production*, 220, pp. 350–357.

Liu, L., Yin, Z. H., Wang, P., Gan, Y. W. and Liao, X. W. (2020). Water-carbon trade-off for inter-provincial electricity transmissions in China. *Journal of Environmental Management*, p. 268.

National Bureau of Statistics of China. (2018). *Chinese statistics yearbook* (in Chinese). Beijing, China: China Statistics Press.

National Energy Administration of China. (2014). *List of twelve electricity transmission corridors for air pollution control to be approved by the national energy administration*. Beijing, China: NEA.

National Energy Administration. (2018). Creating a new record Towards the commanding heights – China's power grid reform, opening up and growth path Observations, *China Electric Power News*.

National People's Congress. (2001). *The tenth five-year plan*. Beijing, China: National People's Congress.

Peng, W., Yuan, J., Zhao, Y., Lin, M., Zhang, Q., Victor, D. G. and Mauzerall, D. L. (2017). Air quality and climate benefits of long-distance electricity transmission in China. *Environmental Research Letters*, 12, p. 064012.

Pfister, S., Koehler, A. and Hellweg, S. (2009). Assessing the environmental impacts of freshwater consumption in LCA. *Environmental Science & Technology*, 43, pp. 4098–4104.

Qu, S., Wang, H., Liang, S. and Shapiro, A. M. (2017). A Quasi-input-output model to improve the estimation of emission factors for purchased electricity from interconnected grids. *Applied Energy*, 200, pp. 249–259.

Su, B. and Ang, B. (2012). Structural decomposition analysis applied to energy and emissions: Some methodological developments. *Energy Economics*, 34, pp. 177–188.

7 Water constraints on alternative energy sources

Climate change impacts on global hydropower potentials

Hydropower production makes significant contributions to global power provision. In 2017, global hydropower production reached 4185 TWh, contributing to 16.4% of global electricity supplies. Total installed hydropower capacity reached 1,267 GW, with 21.9 GW newly added in 2017. China, the world's largest hydropower producer, accounted for nearly half of that newly added capacity, at 9.1 GW. It was followed by Brazil (3.4 GW), India (1.9 GW), Portugal (1.1 GW) and Angola (1.0 GW). The regional distributions of hydropower capacity and hydropower production are shown in Figure 7.1, which shows that the largest amount of hydropower capacity is installed in the East Asian region.

Though hydropower supplies are vulnerable to prolonged droughts, on the whole hydropower is much less constrained by intermittency issues than other renewable supplies, such as solar and wind. Hydropower is also a more flexible electricity source compared to conventional thermal electric power plants. Hydropower stations can be ramped up and down very quickly, normally around a few minutes, to adapt to different levels of electricity demands (Robert, 2010). On the other hand, dam construction has been widely criticized for social, environmental and ecological impacts.

Hydropower production has increased globally but with decreasing proportion in the total electricity supply. Global generation of hydropower has been growing steadily by about 2.3% per year on average since 1980 (Hamududu and Killingtveit, 2012). As shown in Figure 7.2, although globally the share hydropower contributes to the total electricity supply has reduced slightly from 20% to 15% from 1980 to 2015, the sheer amount of hydropower production has increased from just over 2,000 TWh to nearly 4,000 in 2015. Hydropower accounts for the least share in the Middle East

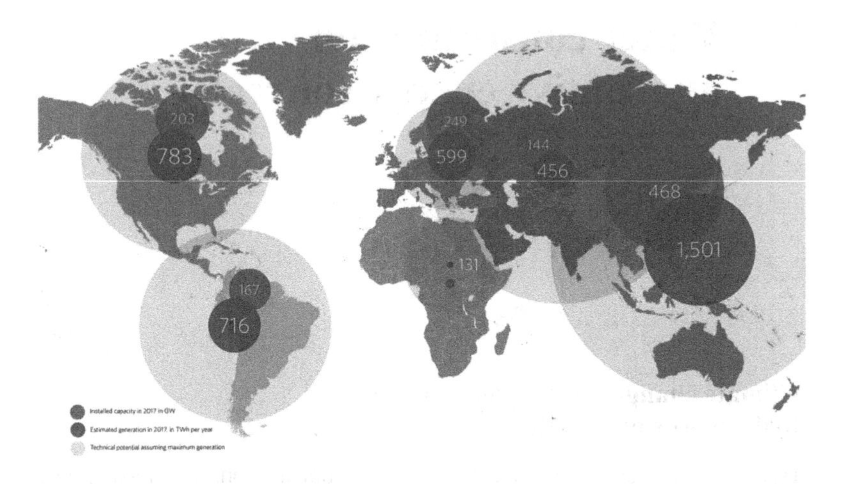

Figure 7.1 Regional distribution of global installed hydropower capacity

Source: International Hydropower Association, 2018.

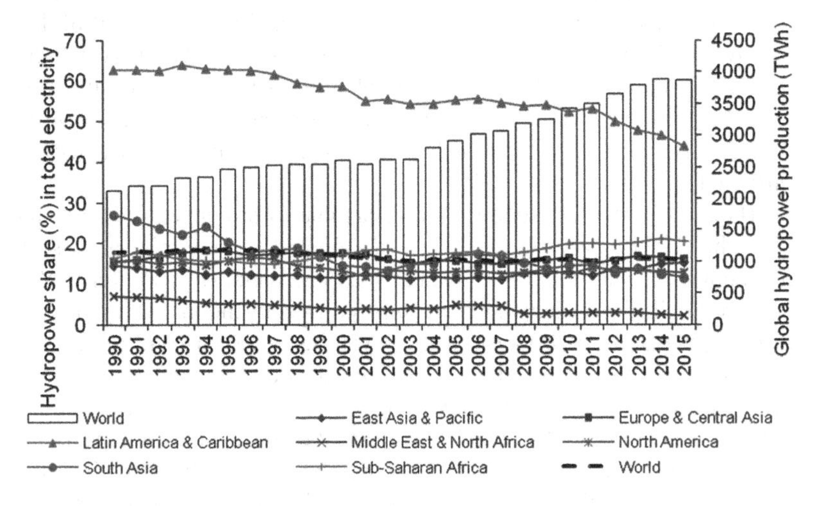

Figure 7.2 Trends of global hydropower productions

Source: Author made based on data from World Bank (2020) and Global Energy Statistic Yearbook 2019.

and North Africa (less than 5%), while it is highest in Latin America and the Caribbean (higher than 40%).

Global hydropower production is expected to continue growing (International Hydropower Association, 2018), though hydropower's share in renewable energy electricity generation is expected to continue falling from 62% in 2018 to 28% in 2050 (Figure 7.3). Nonetheless, hydropower production is still expected to experience a moderate increase. At least 3,700 major dams (> 1 MW installed capacity) are either under construction or in planning across the developing world (Zarfl et al., 2014).

Hydropower production relies on running water in the global river systems. Reduced stream flows can result in a decrease of hydropower production, which is demonstrated by the substantial decreases of hydropower production in California during a prolonged drought period (Gleick, 2015). Van Vliet et al. (2016) calculated that the global average hydropower utilization rates were reduced by 5.2% during drought periods compared to the average from 1981 to 2010. During drought periods, electricity production from conventional fossil fuels is increased to compensate for decreased hydropower production. Herrera-Estrada et al. (2018) calculated that across 11 western states in the US reduction in hydropower production from 2001 to 2015 led to an additional 100 million tons of carbon dioxide emissions for fossil fuel power plants.

Climate change is projected to result in alterations to the spatial and temporal distribution of water availability throughout the world (Arnell, 2004; Van Vliet et al., 2013; Haddeland et al., 2014). Using a Variable Infiltration Capacity model, forced with outputs from five bias-corrected general circulation models, Van Vliet et al. (2016) projected reductions in the usable capacity for 61–74% of 24,515 hydropower plants around the world during the 2050s (2040–2069), compared to 1971–2000 under different emission

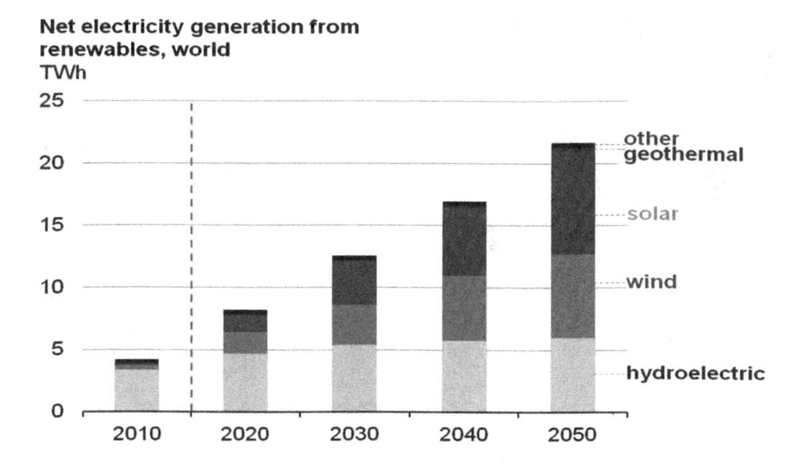

Figure 7.3 Projected global renewable electricity increase to 2050

Source: International Energy Agency, 2019, Global Energy Outlook.

scenarios. Decreases of river discharges are projected for 8% of the global surface area for the 2050s, including in the US, southern and central Europe, Southeast Asia and southern parts of South America, Africa and Australia. Although increased river discharges are also predicted for some regions, such as Canada, northern Europe, Central Africa, India and northeastern China, the current hydropower plants are mostly located in regions where considerable declines in streamflow are projected, resulting in mean reductions in hydropower usable capacity. They found reductions in the global annual hydropower capacities of 1.7–1.9% (2020s), 1.2–3.6% (2050s) and 0.4–6.1% (2080s) based on the GCM ensemble mean for different emission scenarios.

Hydropower production in China

China has the second-largest gross hydropower potential (total energy of natural discharge falling to the lowest level over the entire domain) in the world, only next to Russia (Zhou et al., 2015). According to a survey by

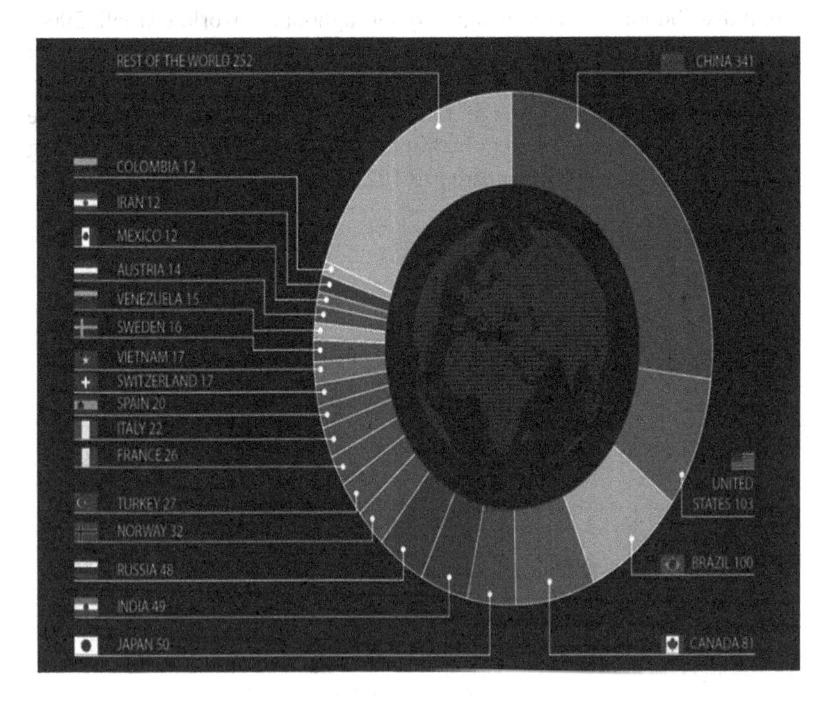

Figure 7.4 Hydropower installed capacity (GW) of top 20 countries including pumped storage in 2017

Source: International Hydropower Association, 2018.

Yan et al. (2006), China's hydropower potential and technical exploitable capacity are respectively 694 and 542 GW. According to IHA (2018), by 2017, China had installed 341 GW of hydropower (352 GW by 2018), which topped the global list and surpassed the combined total of the three countries that followed (i.e. US (103 GW), Brazil (100 GW) and Canada (81 GW)) (Figure 7.4).

As can be seen from Figure 7.5, hydropower accounts for a consistent 15% to 20% of China's entire electricity portfolio, with the amount of hydropower production having increased from just above 200 TWh in 2000 to just below 1200 TWh in 2017. The trends of hydropower productions also corresponded with drought incidents. For instance, hydropower production plummeted in 2011 during the prolonged droughts in the lower reach of the Yangtze River (News Weather, 2011).

China's hydropower capacity is distributed unevenly across the country. Most reservoirs are located in southern regions where water is abundant, and a few are in north China, mostly in the Gansu and Qinghai provinces at the upper reach of the Yellow River. By the end of 2018, the top five provinces with the largest hydropower capacity in China were Sichuan (78.24 GW), Yunnan (66.66 GW), Hubei (36.75 GW), Guizhou (22.12 GW) and Guangxi (16.75 GW). In 2018, Sichuan, Yunnan and Hubei generated over 315.69, 269.85 and 148.41 TWh hydroelectricity respectively. To put it in context, hydropower production in Sichuan alone is enough to power the United Kingdom for an entire year.

Different provinces rely on hydropower productions to different extents. As can be seen from Figure 7.7, hydropower contributes to more than 80% of the total electricity production in Tibet, Sichuan and Yunnan. Final

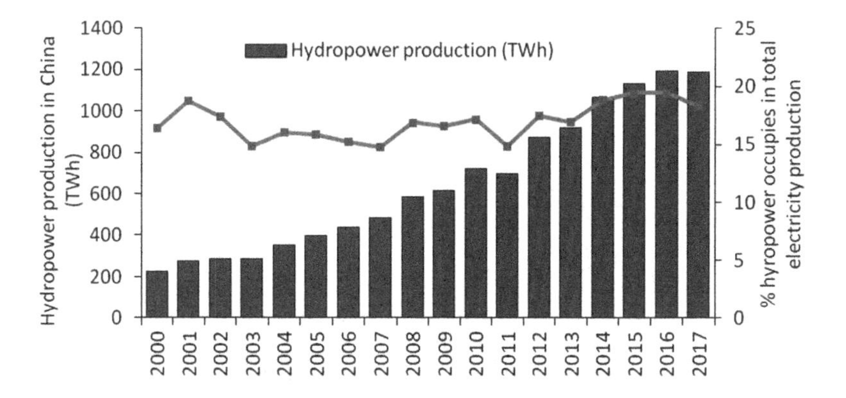

Figure 7.5 Hydropower productions in China

Source: National Statistic Bureau (2018).

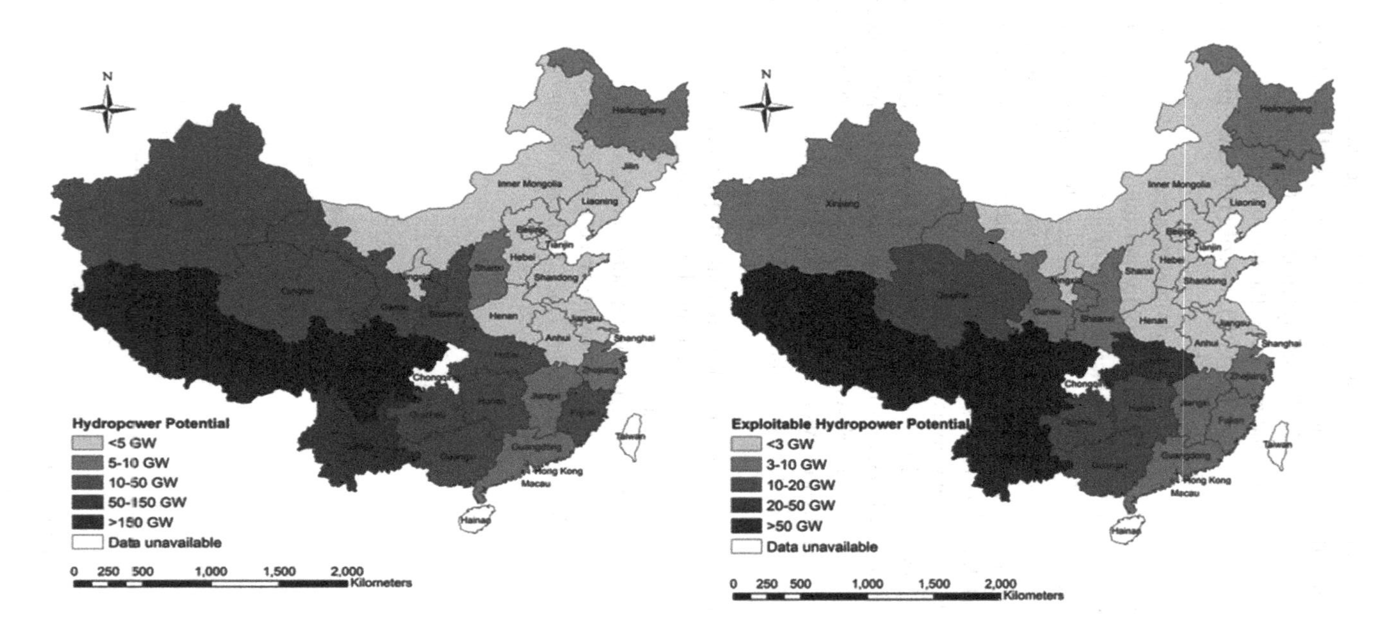

Figure 7.6 Distribution of hydropower capacity in China

Source: National Statistic Bureau (2018).

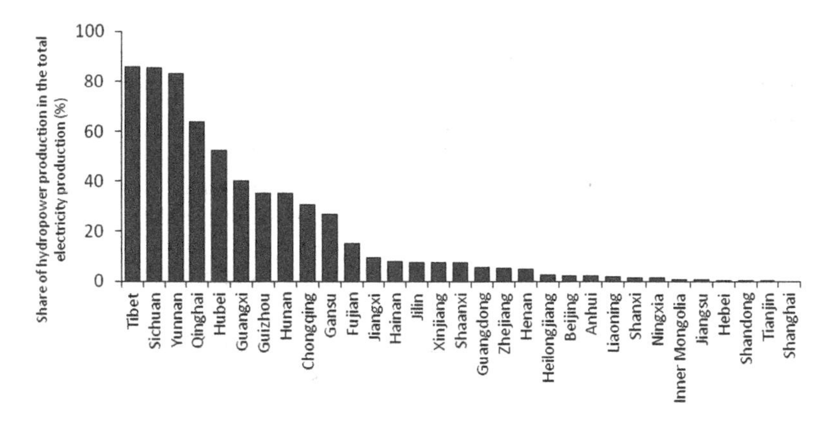

Figure 7.7 Share of hydropower production in provincial total electricity production in 2018

Source: National Statistic Bureau (2018).

electricity consumption in these provinces heavily relies on sufficient flow in the river channels. Moreover, as mentioned in the previous chapter, the 'West-to-East' electricity transmission lines transfer large amounts of hydropower produced in the southwestern and central provinces (i.e. Sichuan, Yunnan and Hubei) to support electricity demands in the eastern coastal provinces, such as Guangdong, Shanghai and so forth.

China has committed to peak its CO_2 emissions by 2030. In order to meet its intended national determined contribution, China is planning to further expand its hydropower production. By 2020, China plans to install a total number of 385 GW of hydropower capacity, including 45 GW of pumped storage. By 2030, the numbers are expected to increase to 570 GW (total hydropower capacity), including 120 GW of pumped storage. By 2030, non-fossil energy is expected to account for 20% of the primary energy mix in China.

Projected impact of climate change on hydropower production in China

Hydropower productions consume large amounts of water due to evaporation in dammed reservoirs. It accounts for the predominant share of water consumption for power productions in China. From 2002 to 2010, hydropower's water consumption in China grew substantially from 6.1 to 14.6 billion m^3, which represented some 80% of the power sector's total water consumption (Liao et al., 2018). As described in Chapter 2, water consumption at reservoirs differs substantially in different places, depending on local geographical, hydrological and climatic conditions. According to Liu et al. (2015), average water consumption intensity for hydropower production in China amounts to 12.86 m^3 per MWh of electricity produced. As shown in Figure 7.8, Liu et al. (2015) examined 209 hydropower stations in China, and their water consumption intensity ranges from 0.0036 to 15121.43 m^3/MWh. There is a general trend of relative lower water consumption intensity at upstream and larger water consumption intensity at downstream within a river basin. Hydropower production's water consumption intensity is particularly high in the arid and semi-arid northern regions where local climatic and land surface conditions (e.g. high wind and little vegetation) contribute to high evaporation rates in reservoirs.

It should be noted that although hydropower stations have much larger consumptive water losses than coal power plants, water consumption for coal power productions should not be overlooked since their impacts on water resources are site-specific and depend on local water scarcity levels. While large hydropower stations are often developed at large waterways because they depend on sufficient flows to generate electricity, coal power plants are often built close to coal mines to save coal transportation cost

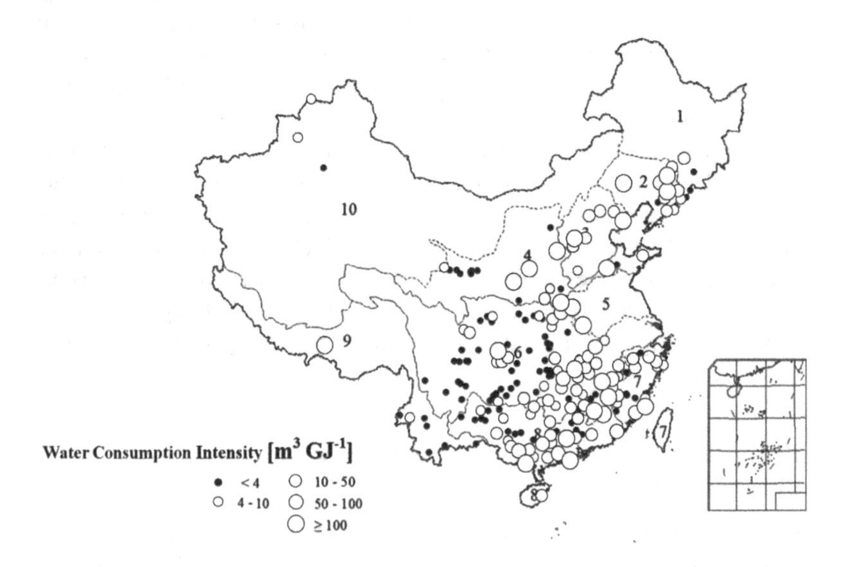

Figure 7.8 Water consumption intensity for hydropower production in China's 209 hydropower stations

Source: Liu et al., 2015.

and are much less concerned about water availability or impacts on water resources. Since China's coal reserves are mostly endowed in water-stressed areas, for instance, the Yellow River Basin is endowed with merely 2% of China's renewable water resources, but with almost 50% of the country's total coal reserves, growing coal power generation imposes increasing pressures on scarce water resources. Furthermore, many coal power plants turn to groundwater resources if sufficient surface water supplies are not available, which leads to deteriorating groundwater over-exploitation issues.

Based on multi-model hydrological projections by the Inter-Sectoral Impact Model Intercomparison Project (ISI-MIP) (Warszawski et al., 2014), Liu et al. (2016) investigated the impacts of future climate change on China's gross hydropower potential as well as on the existing hydropower facilities. They found that the gross hydropower potential of China is projected to largely decrease in southwestern China and south-central China, especially in the summer period, while increasing in most areas in northern China. However, because the existing hydropower facilities are mostly located in southwestern and south-central China, the usable capacity is expected to decrease by about 2.2% to 5.4% from 2020 to 2050 and decrease by 1.3% to 4% from 2070 to 2099. Their results are consistent with the global findings by Van Vliet et al. (2016). Despite that global hydropower potential is projected to increase, the usable capacity is expected to decrease because most existing facilities are located in regions where reduced water availability is projected.

Water constraints on concentrated solar power

Besides hydropower, potential water limits may pose constraints on the large deployment of concentrated solar power (CSP). CSP typically uses large arrays of ground mirrors to concentrate sunlight to transfer heat through a medium (e.g. oil, water or salt) and then to generate steam to spin a turbine to generate electricity. CPS technology uses water in the steam cycle and cooling process. Moreover, CSP also requires water for cleaning the mirrors, especially in dry areas. As illustrated in Figure 7.9, Macknick et al. (2012) have conducted a review study of electric power plants in the US and consolidated the values of water consumption of different generating technologies and energy sources. CSP with tower cooling technologies require similar amounts of water inputs as coal power plants per unit of electricity generated.

It can also be seen from Figure 7.9 that theoretically dry cooling technology is able to reduce water use for CSP significantly. However, as air has a much lower capacity to dissipate residual heat than water, a large number of cooling fans need to be deployed, which results in large amounts of

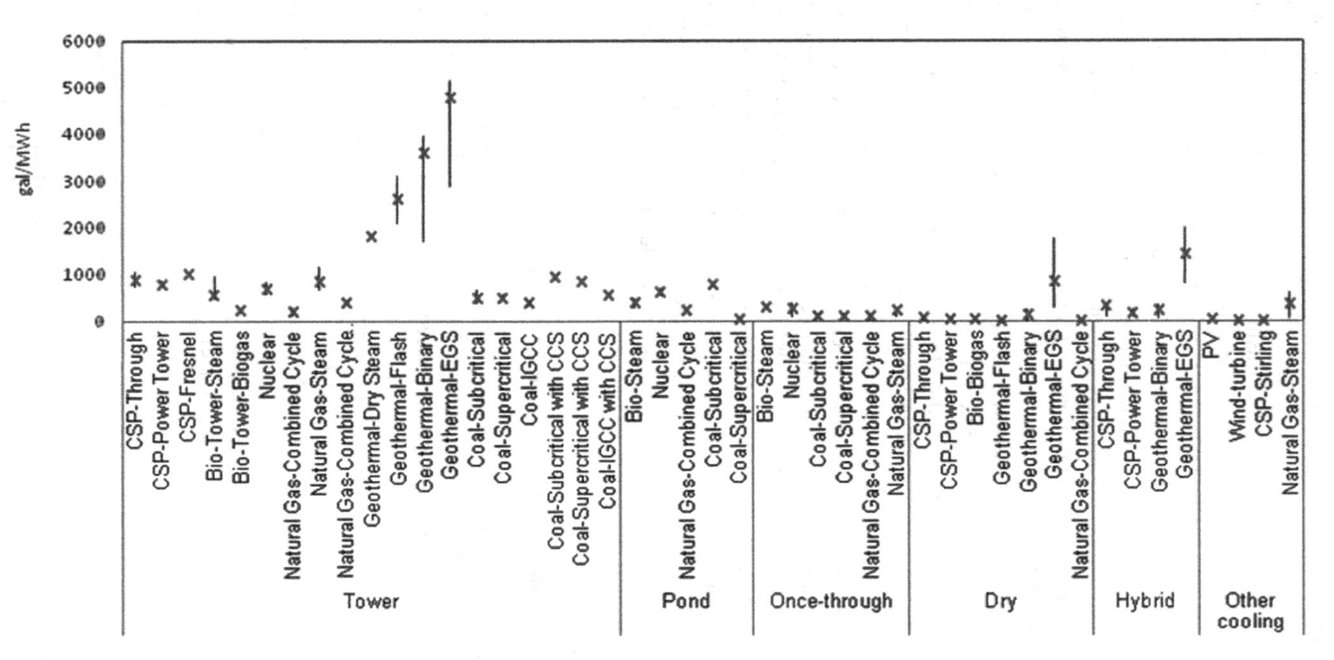

Figure 7.9 Operational water consumption factors for different types of electricity productions

Source: Macknick et al., 2012.

Note: IGCC: Integrated Gasification Combined Cycle; CCS: Carbon capture and sequestration; CSP: Concentrated solar power. Upper and lower ends represent maxima and minima, respectively; Red Cross represents median values.

auxiliary electricity demand with reduced conversion efficiency. Especially during the summer, while electricity demand is particularly high, CSP using dry cooling technology has low thermal efficiency due to the high air temperature. As a result, researchers were unable to identify any large-scale CSP facilities around the world that use dry cooling systems (Carter and Campbell, 2009).

China's CSP development is in the early stage but is projected to grow rapidly. In 2011, China's National Development and Reform Commission issued an 'Industry Structure Adjustment Catalogue', which has given high priority to CSP development as one of the newly encouraged energy resources. China's first MW-class CSP power plant, 'Dahan', entered into operation in 2012. China National Energy Administration launched the first batch of CSP pilot projects in 2016 including 20 projects with a total capacity of 1.35 GW. According to IEA estimates (2018), China is expected to overtake the US to have the world's second-largest CSP installed capacity by 2023, with 1.9 GW coming online, following 2.3 GW in Spain. From 2017 to 2018, China has further completed a feasibility study for another 24 projects with a total capacity of 3.05 GW. In 2019, 57 projects with a total capacity of 14.9 GW have been submitted for the application of China's second batch of CSP demonstration projects (CSP Focus, 2019).

China's most promising CSP sites are located in dry northern regions where water availability is low. As can be seen from Figure 7.10, there are mismatches between China's water resources (Tu et al., 2016) and potential solar power locations. While China's solar energy sites are mostly endowed in the northwestern regions, water is less abundant in those regions. The largest five provinces for CSP development are Qinghai, Gansu, Hebei, Inner Mongolia and Xinjiang, which have made plans for developing CSP to, respectively, 20, 5.6, 6, 16 and 20 GW by 2030. Among these provinces, in 2018, Hebei's annual water withdrawal (18.24 km^3) already exceeded its annual water availability (13.83 km^3). The water Withdrawal-to-Availability (WTA) ratio in the other four provinces are 0.63 (Xinjiang), 0.42 (Inner Mongolia), 0.34 (Gansu) and 0.02 (Qinghai) (National Statistic Bureau, 2019). Except Qinghai, all of the other four provinces are facing different levels of water stress. Future CSP development may face potential water constraints in those provinces, which needs to be further studied.

Nuclear power plants

Water scarcity also affects cooling for inland nuclear power plants. Figure 7.9 shows that nuclear power plants require the largest water inputs to produce the same amount of electricity due to the comparatively low energy conversion efficiency. Reliable water supplies are crucial for the sake of

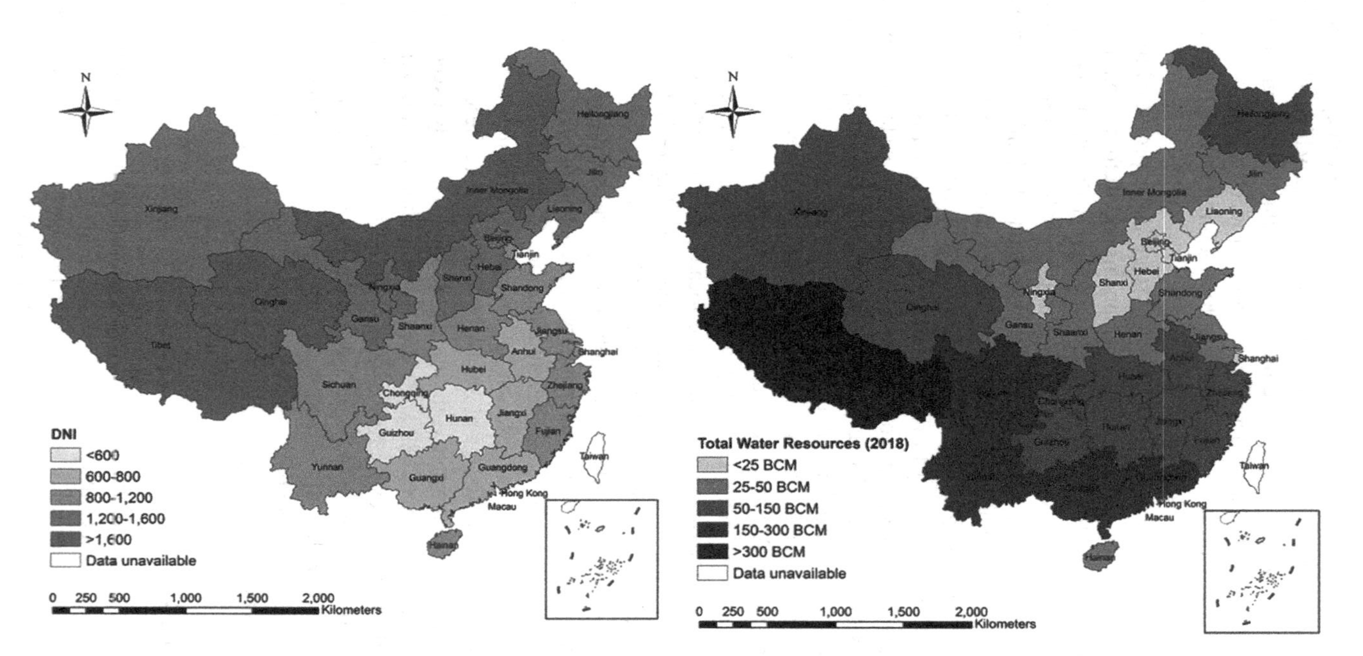

Figure 7 10 China's provincial average Direct Normal Irradiation (DNI) (left) and total water resources

nuclear security. While China's current nuclear reactors are all located on coastal regions and use seawater for cooling purposes, development of inland nuclear power plants has been discussed (World Nuclear Association, 2015). Site selections for potential nuclear power plants require comprehensive examination, and relevant information is not available to us. However, it should be highlighted that water availability assessment incorporating the impacts of future climate change must be included as a key criterion for its feasibility studies.

References

Arnell, N. W. (2004). Climate change and global water resources: SRES emissions and socioeconomic scenarios. *Global Environmental Change*, 14(1), pp. 31–52.

Carter, N. T. and Campbell, R. J. (2009). *Water issues of concentrating solar power (CSP) electricity in the U.S. Southwest*. Washington, DC: Congressional Research Service.

CSP Focus. (2019). *Applicants in China 2nd batch of CSP demonstration projects*. Beijing, China: CSP Focus.

Gleick, P. H. (2015). *Impacts of California's ongoing drought: Hydroelectricity generation*. Oakland: Pacific Institute.

Haddeland, I., Heinke, J., Biemans, H., Eisner, S., Flörke, M., Hanasaki, N., Konzmann, M., Ludwig, F., Masaki, Y., Schewe, J. and Stacke, T. (2014). Global water resources affected by human interventions and climate change. *Proceedings of the National Academy of Sciences*, 111(9), pp. 3251–3256.

Hamududu, B. and Killingtveit, A. (2012). Assessing climate change impacts on global hydropower. *Energies*, 5, pp. 305–322.

Herrera-Estrada, J. E., Diffenbaugh, N. S., Wagner, F., Craft, A. and Sheffield, J. (2018). Response of electricity sector air pollution emissions to drought conditions in the Western United States. *Environmental Research Letters*, 13(12), p. 124032.

Huggins, R. A. (2010). *Energy storage*. New York: Springer, p. 60.

International Energy Agency. (2019). *Global energy outlook*. Paris, France: IEA.

International Hydropower Association. (2018). *2018 hydropower status report*. London: International Hydropower Association.

Liao, X., Zhao, X., Hall, J. W. and Guan, D. (2018). Categorising virtual water transfers through China's electric power sector. *Applied Energy*, 116, pp. 252–260.

Liu, J., Zhao, D., Gerbens-Leenes, P. W. and Guan, D. (2015). China's rising hydropower demand challenges water sector. *Scientific Reports*, 5, p. 11446.

Liu, X., Tang, Q., Voisin, N. and Cui, H. (2016). Projected impacts of climate change on hydropower potentials in China. *Hydrology and Earth System Sciences*, 20, pp. 3343–3359.

Macknick, J., Newmark, R., Heath, G. and Hallett, K. C. (2012). Operational water consumption and withdrawal factors for electricity generating technologies: A review of existing literature. *Environmental Research Letters*, 7, p. 045802.

National Development and Reform Commission. (2011). *Industry structure adjustment catalogue* (in Chinese). Beijing, China: NDRC.

National Energy Administration. (2016). *Notice on construction of concentrated solar power demonstration projects* (in Chinese). Beijing, China: NEA.

National Statistic Bureau. (2018). *National Statistic Yearbook 2018*. Beijing, China.

National Statistic Bureau. (2019). *Statistic yearbook*. Beijing, China: National Statistic Bureau.

News Weather. (2011). *Why the lower reach of Yangtze encounters rare droughts?* (in Chinese). Available at: http://news.weather.com.cn/1348984.shtml [Accessed 27 Jan. 2020].

Tu, M., Wang, F., Zhou, Y. and Wang, S. (2016). Gridded water resource distribution simulation for China based on third-order basin data from 2002. *Sustainability*, 8(12), p. 1309.

Van Vliet, M. T. H., Franssen, W. H., Yearsley, J. R., Ludwig, F., Haddeland, I., Lettenmaier, D. P. and Kabat, P. (2013). Global river discharge and water temperature under climate change. *Global Environmental Change*, 23(2), pp. 450–464.

Van Vliet, M. T. H., Wiberg, D., Leduc, S. and Riahi, K. (2016). Power-generation system vulnerability and adaptation to changes in climate and water resources. *Nature Climate Change*, 6, pp. 375–381.

Warszawski, L., Frieler, K., Huber, V., Piontek, F., Serdeczny, O. and Schewe, J. (2014). The inter-sectoral impact model intercomparison project (ISI – MIP): Project framework. *Proceedings of the National Academy of Sciences of the United States of America*, 111, pp. 3228–3232. doi:10.1073/pnas.1312330110.

World Bank. (2020). *World Bank open data*. Washington, DC: World Bank.

World Nuclear Association. (2015). *Nuclear power in China*. Available at: www.worldnuclear.org/info/country-profiles/countries-a-f/china-nuclear-power.

Yan, Z., Peng, C., Yuan, D. and Qian, G. (2006). General description of the re-check work of the national hydropower resources survey (in Chinese). *Water Power*, 32, pp. 8–11.

Zarfl, C., Lumsdon, AE., Berlekamp, J., Tydecks, L. and Tockner, K. (2014). A global boom in hydropower dam construction. *Aquatic Sciences*, 77, pp. 161–170.

Zhou, Y., Hejazi, M., Smith, S., Edmonds, J., Li, H., Clarke, L., Calvin, K. and Thomson, A. (2015). A comprehensive view of global potential for hydro-generated electricity. *Energy & Environmental Science*, 8, pp. 2622–2633.

8 Policy interactions between water and electricity sectors in China

Energy policies aimed at improving water management at coal power plants

Policies in the electric power sector can generate either direct or indirect impacts on water resources. Sanders (2015) identified five aspects of the power sector that have impacts on water resources: (1) fuel mix; (2) cooling technology configuration; (3) relevant regulations; (4) changing climate; and (5) power grid characteristics, including transmission and so on. Policies that result in impacts on any one of those five aspects could affect the power sector's total water uses. Webster, Donohoo and Palmintier (2013) found that a restriction on CO_2 emissions in the energy sector in Texas also reduces water withdrawal by its power sector; Bartos and Chester (2014) have shown that Arizona's Energy Efficiency Mandate and Renewable Portfolio Standard have resulted in considerable water savings. China has issued a series of policies directly aimed at improving water performance at coal power plants, including (1) setting requirements on the maximum water withdrawal intensities for coal power plants; (2) banning groundwater usage in water-scarce regions, particularly in groundwater over-exploited regions; (3) forcing the deployment of air-cooling technologies; (4) encouraging the utilization of other unconventional water sources, including reclaimed water from municipal wastewater, coal mine drainage and so forth.

There are various mandatory water withdrawal standards for coal power plants. At a national level, a new national water withdrawal standard (GB/T 18916.1–2012) replaced 'GB/T 18916.1–2002' in 2012 and set more stringent requirements regarding water uses by coal power plants with closed-loop cooling systems (General Administration of Quality Supervision, Inspection and Quarantine of China, 2002, 2012). It is noteworthy that air-cooling units are not included in the national standard, while it is regulated in some regions, as shown in Table 8.1. It should be noted that: (1) Regulations on water uses by power plants equipped with air-cooling

Table 8.1 Provincial coal power plants' water use standards in China

Province	*year*	*Unit (m^3/ MWh if not noted)*	*Standard*	*Capacity (MW)*	*Cooling system*	*Reference*
Hunan*	2008		5	<300	closed-loop	GB/T18916.1–2002
			4	≥300		
			130	<300	open-loop	
			120	≥300		
Liaoning	2008		1	<300	open-loop	
			0.65	≥300		
			1.2	<300	air	
			0.8	≥300		
Hebei	2009	m^3/MWh	3	<300	closed-loop	GB/T18916.1–2002
			2.39	≥300		
			2.15	≥600		
			1.2	<300	open-loop	
			0.72	≥300		
			1.2	<300	air	
			0.96	≥300		
			0.86	≥600		
		m^3/S^*GW	1	<300	closed-loop	GB/T18916.1–2002
			0.8	≥300		
			0.75	≥600		
			0.2	<300	open-loop	
			0.12	≥300		
			0.3	<300	air	
			0.2	≥300		
			0.16	≥600		
Inner Mongolia	2009	m^3/MWh	4.8	<300	closed-loop	GB/T18916.1–2002
			3.84	≥300		
			1.2	<300	open-loop	
			0.72	≥300		
			0.8		air	
		m^3/S^*GW	1	<300	closed-loop	GB/T18916.1–2002
			0.8	≥300		
			0.2	<300	open-loop	
			0.12	≥300		
			0.18		air	
Qinghai	2009	m^3/MWh	4.8	<300	closed-loop	GB/T18916.1–2002
			3.84	≥300		
			1.2	<300	open-loop	
			0.72	≥300		

Province	*year*	*Unit (m^3/MWh if not noted)*	*Standard*	*Capacity (MW)*	*Cooling system*	*Reference*
		m^3/S*GW	1	<300	closed-loop	GB/T18916.1–2002
			0.8	≥300		
			0.2	<300	open-loop	
			0.12	≥300		
Henan	2009		4.8	<300	closed-loop	GB/T18916.1–2002
Jilin	2009		3.84	≥300		
Sichuan	2010		1.2	<300	open-loop	
Gansu	2011		0.72	≥300		
Shandong	2009		3	<300	closed-loop	
			2.5	≥300		
			1	<150	open-loop seawater	
			0.7	≥150		
			0.5	≥300		
Shaanxi	2010		1.2	≥300	air	
			1.5	<300		
			4	≥300	closed-loop	
			5	<300		
Guizhou	2011		4.2	<150		
			3.6	≥150		
			2.9	≥300		
			2.8	≥600		
			2.7	≥1000		
Jiangxi*	2011		120	<600	open-loop	
			100	≥600		
			4.8	<300	closed-loop	
			3.84	≥300		
Anhui*	2013		60–70	≥300	open-loop	
			100–110	<300		
			3.2	<300	closed-loop	GB/T18916.1–2012
			2.75	≥300		
			2.4	≥600		
Yunnan	2013		3.2	<300	closed-loop	
			2.75	≥300		
			2.4	≥600		
			0.79	<300	open-loop	
			0.54	≥300		
			0.46	≥600		
			0.95	<300	air	
			0.63	≥300		
			0.53	≥600		

Note: *denotes the provinces where water withdrawal of open-loop cooling systems is controlled instead of water consumption as in other provinces.

systems are mostly in place in the north, including Liaoning, Hebei, Inner Mongolia and Shandong; (2) provincial standards formulated later are more stringent as they conform to the more recent national standard; (3) two sets of standards – m^3/MWh and m^3/S*GW – existed in earlier regulations (Hebei, Qinghai and Inner Mongolia), which were designed for assessments for operation and commission respectively, as m^3/MWh is generally more stringent. Later on, only m^3/MWh is adopted.

There are also restrictions on groundwater usage and air-cooling deployment. In water-scarce areas, new coal power plants are not allowed to extract groundwater, especially in groundwater over-exploited regions since 2004 (National Development and Reform Commission, 2004). Although air-cooling units are not included in the national water withdrawal standards, NDRC (2004) required new coal power plants in water-scarce northern regions to employ large air-cooling units with water withdrawal intensities of less than 0.18 m^3/(s GW). The Ministry of Water Resources (MWR) also issued a similar requirement in 2013 (MWR, 2013), which required new coal power plants in water-stressed areas to deploy air-cooling systems with water withdrawal intensities lower than 0.1 m^3/(s GW). The MWR requirement is stricter than the NDRC requirement and even stricter than the national water withdrawal standard issued later that year that sets the water withdrawal limit for air-cooled units larger than 500 MW at 0.13 m^3/(s GW). As a result, new power plants equipped with air-cooling systems need much less time to obtain environmental permits and thus have proliferated during the last decade. Nationally, the percentage of air-cooling systems has increased from 6.4% in 2000 to 14.5% in 2015 while that of wet cooling systems, including both closed-loop and open-loop cooling, has decreased accordingly.

Coal power plants are encouraged to use unconventional water resources. China's 12th Five-Year Plan (2011–2015) was supposed to develop 14 coal bases, which are all located in the northern dry regions. In response to such planning, MWR issued 'Regulations on implementing water resources evaluation for the planning and development of large coal power industrial clusters' in 2013, requiring newly constructed coal power plants in northern China to prioritize reclaimed water and coal mine drainage. For example, in the water-stressed Yellow River Basin, Ministry of Water Resources allocated 417.6 million m^3 of water withdrawal quotas for 84 GW of newly added coal power plants from November 2009 to December 2014, among which 58.3% (243.2 million m^3) are reclaimed water (Zhang et al., 2014). In the country's second-largest coal producer Shanxi province, it is required that the coal mine drainage recycling rate should be increased from the current 67.7% to 75%.

Other governmental agencies have also issued technical guidelines to promote water-saving practices in coal power plants. Those official but nonmandatory guidelines for promoting water-saving practices in the coal power sector include (1) 'Guidelines for water saving in thermal power plants' issued by the State Economic and Trade Committee (2001), (2) 'China water conservation technology policy outline' jointly issued by NDRC and several other relevant ministries in 2005, and (3) 'Water efficiency guide for key industrial sectors' jointly issued by the Ministry of Industry and Information Technology and several other relevant ministries in 2013.

Energy policies that generate co-benefits of or tradeoffs with water savings

Improving energy efficiency in coal power plants can generate co-benefits of water saving. China's 8th Five-Year Plan (1991–1995) imposed an efficiency requirement on coal power plants (National People's Congress 1991). In 2007, in order to lower energy consumption and pollution, China's State Council issued a 'Notice on closing down small thermal electric power generating units'. It required new power plants of large capacity to be built on the condition that an equivalent number of small units were shut down. As a result, during the period of 2006–2011, with China's 11th Five-Year Plan (National People's Congress 2006), a total of 77 GW of small units were replaced by large supercritical (600 MW) and ultra-supercritical (1,000 MW) units (Wu and Huo, 2014). The graph that follows (Figure 8.1)

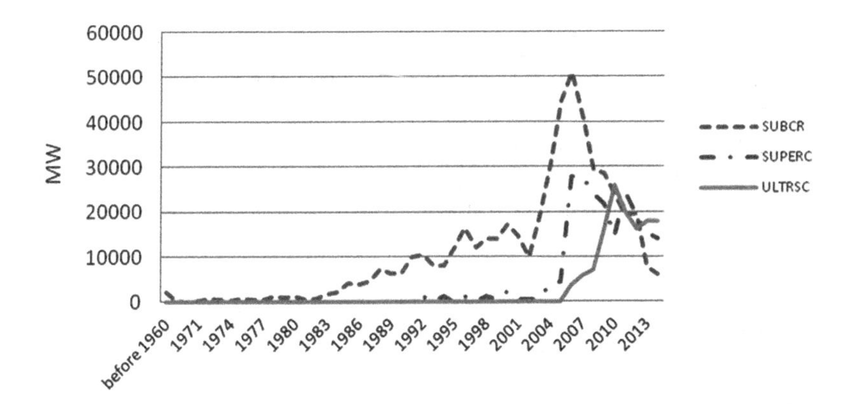

Figure 8.1 Coal power plants with different boiler technologies built every year in China

Note: (SUBCR – subcritical; SUPERC – supercritical; ULTRSC – ultra-supercritical).

indicates that the construction of subcritical power plants peaked in 2007 when the 'Notice on closing down small thermal electric power generating units' was issued. Since 2012, the yearly construction of ultra-supercritical power plants started to surpass the two other kinds. On the contrary, that of subcritical units has fallen to the lowest at around 6,000 MW in 2014.

Larger electricity-generating units have higher efficiency and lower water intensities. As Jiang and Ramaswami (2015) have demonstrated based on field data in China, electricity-generating units with larger capacities are more water efficient. The reason is because larger units tend to adopt more advanced boiler technologies (e.g. supercritical, with higher efficiencies) so that more heat is converted from the primary energy carrier (e.g. coal) to electricity, and therefore less residual heat needs to be dissipated by cooling water. The relationship between boiler technology and unit capacity can be seen in Table 8.2. Units with larger capacities tend to employ more advanced technologies which have higher energy efficiencies and better performance in many regards (e.g. emission, water consumption).

Policies aimed at reducing negative environmental impacts from the coal power sector can also generate tradeoffs with water conservation. Coal power production has also given rise to a number of different negative environmental impacts, such as the emission of sulfur dioxide, a pollutant gas that has contributed to acid rain and caused serious detrimental health impacts (Shi et al., 2017). China has mandated all coal power plants to install desulfurization facilities. The most predominant desulfurization process adopted is called the wet flue gas desulfurization process, where water is often used as a solution to absorb sulfur dioxide. Nationally, desulfurization in China's coal-fired power plants removed 23.15 million tons of SO_2 throughout the country in 2015 but at the expense of 0.80 $km^3$of additional water consumption (Liao et al., under preparation).

Energy transition to low-carbon renewable energies has generated co-benefits of water saving. As the world's biggest CO_2 emitter, China pledged to increase the share of its non-fossil fuel to around 15% by 2020 at Copenhagen in 2009.

Table 8.2 Boiler technology composition of power plants of different capacities

Capacity (MW)	*High pressure*	*Ultra-high pressure*	*Subcritical*	*Supercritical*	*Ultra-supercritical*
<300	5%	95%			
300–600			96.2%	3.8%	
600–1000			38%	50%	12%
1000					100%

It first proposed to diversify the energy portfolio in its 10th Five-Year Plan (2001–2005) (National People's Congress of China, 2001), and the 12th Five-Year Plan (2011–2015) (National People's Congress of China, 2011) made more specific targets aimed at decreasing the share of coal in the country's energy structure to 63% and increasing that of natural gas from 3.9% to 8.3%. Moreover, China's latest 'Intended Nationally Determined Contribution' (National Development and Reform Commission, 2015) made a new commitment to increase its share of non-fossil fuels in primary energy consumption to around 20% by 2030. China needs to deploy 800 to 1,000 GW in non-fossil fuel capacity, close to the US's total current electricity capacity, to realize that goal (WRI, 2015). In 2016, China halted the construction of more than 100 coal-fired power plants across the country, with a combined output of 100 GW (Greenpeace, 2017), because wind power and solar PV are the two main low-carbon technologies being increasingly adopted to replace coal and they only require negligible amounts of water inputs during the production phase, mainly for cleaning purposes. Such transition in China – the change of the power sector's energy portfolio has generated co-benefits of water saving as described in Chapter 4. Furthermore, China is also looking to develop its floating solar PV facilities. For instance, in 2017, the largest floating solar PV facility (40 MW) in the world at that time entered into operation in the Anhui province in eastern China (World Economic Forum, 2017), which offers multiple co-benefits such as reducing water evaporation from dammed water, increasing energy conversion efficiency, increasing potentials for fishery and so forth.

Water sector policies also have impacts on coal power plants

The Chinese government issued its 'Most Stringent Water Management Mechanism' named the 'Three Red Lines' policy, setting targets on total water use, industrial and agricultural water efficiency and water quality improvements. The first 'Red Line' sets a cap on total water withdrawal at 635, 670 and 700 km^3 in 2015, 2020 and 2030, respectively. Qin et al. (2015) pointed out that this Red Line could potentially be violated by China's electric power sector unless technological innovations are facilitated. Liao et al. (2016) highlighted that such potential conflicts are particularly pronounced in the eastern, central and northern regions.

The second Red Line sets targets for improvements in industrial water efficiency. China's industrial water efficiency is low compared to global standards (World Bank, under preparation). Liao and Ming (2019) assessed the compliance of China's 30 mainland provinces (except Tibet due to data limitations) in meeting their industrial water efficiency improvement targets set by the second Red Line. They found that nine provinces failed to meet

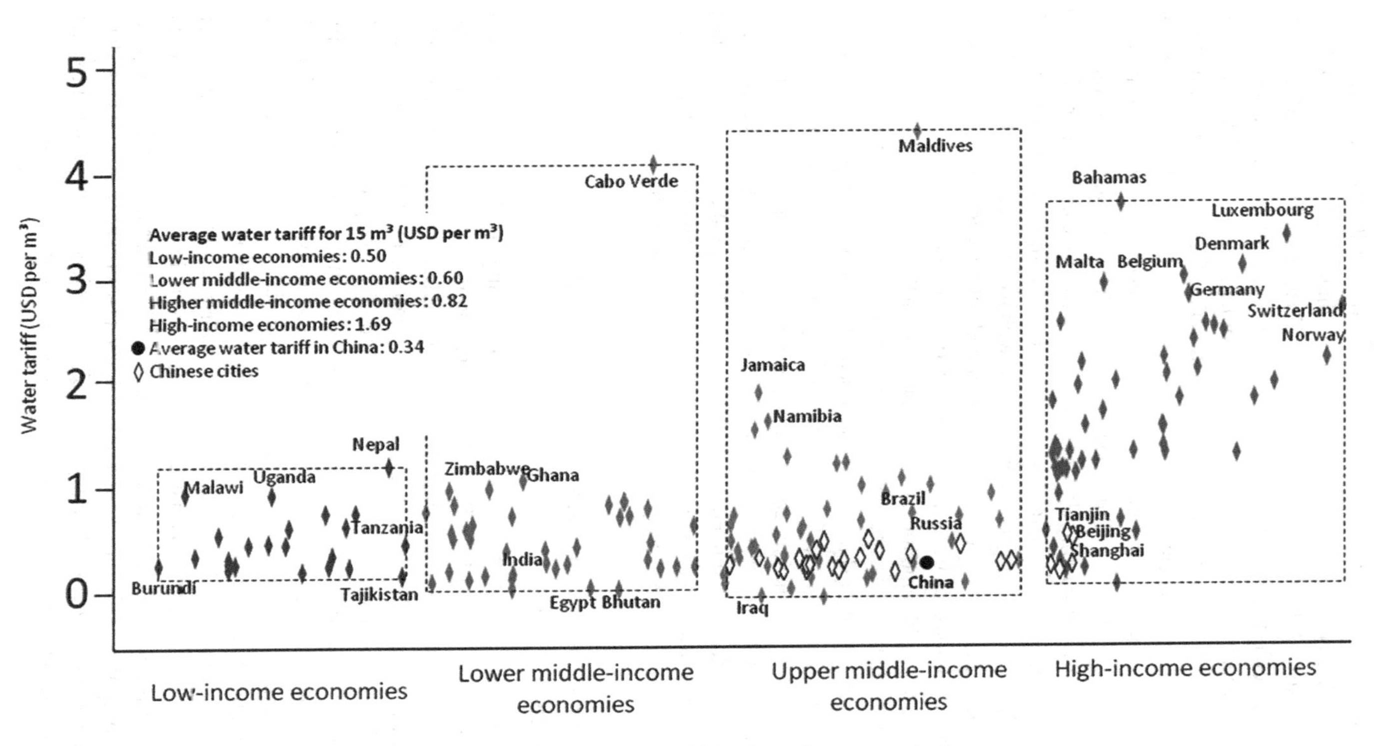

Figure 8.2 China's water tariffs benchmarked with global tariffs (World Bank, under preparation)

their target in 2015, among which six provinces consumed large amounts of scarce water resources for energy productions, including Shanxi, Shandong, Hebei, Henan, Jiangsu and Xinjiang. Improving water efficiency in their energy sectors is an effective way for those provinces to meet their industrial water efficiency improvement target.

Last but not least, China's water tariffs are low compared to the global standards, which do not incentivize water-saving technologies or practices at power plants. It is widely recognized that pricing water resources can provide incentives to improve the efficiency of water use (OECD, 2010; Garrick, Hanemann and Hepburn, 2020). Although it is difficult to obtain worldwide data for water tariffs for industrial uses, which include coal power plants, we have compared global water tariffs for residential uses for reference. Water tariffs in Chinese cities are from China Water Net (h20-china.com), while global water tariffs are from The International Benchmarking Network. As can be seen in Figure 8.2, the highest water tariffs in China's several most water-scarce cities (e.g. Beijing and Tianjin) are on par with average water tariffs in low-income economies. Based on the calculations by Zhang et al. (2016), the cost for water resources accounts for roughly only 1% of the electricity tariff at which coal power plants are selling electricity to the grids, which does not provide enough incentives for coal power plants to adopt water-saving technologies.

References

Bartos, M. D. and Chester, M. V. (2014). The conservation nexus: Valuing interdependent water and energy savings in Arizona. *Environmental Science & Technology*, 48, pp. 2139–2149.

China Water Net. (2020). Available at: h20-china.com.

Garrick, D., Hanemann, M. and Hepburn, C. (2020). Rethinking the economics of water: An assessment. *Oxford Review of Economic Policy*, 36(1), pp. 1–23.

General Administration of Quality Supervision, Inspection and Quarantine of China. (2002). *Norm of water intake: I. Electric power production GB/T 18916.1–2002* (in Chinese). Beijing, China: General Administration of Quality Supervision, Inspection and Quarantine of China.

General Administration of Quality Supervision, Inspection and Quarantine of China, Standardization Administration of China. (2012). *Norm of water intake: I. Fossil fired power production GB/T 18916.1–2012* (in Chinese). Beijing, China: General Administration of Quality Supervision, Inspection and Quarantine of China, Standardization Administration of China.

Greenpeace. (2017). *China halts more than 150 coal-fired power plants*. Available at: https://unearthed.greenpeace.org/2017/10/11/china-halts-150-coal-fired-power-plants/.

The International Benchmarking Network. (2020). Available at: www.ib-net.org/.

Jiang, D. and Ramaswami, A. (2015). The 'thirsty' water-electricity nexus: Field data on the scale and seasonality of thermoelectric power generation's water intensity in China. *Environmental Research Letters*, 10(2), p. 024015.

Liao, X. W., Hall, J. W. and Eyre, N. (2016). Water use in Chinas thermoelectric power sector. *Global Environmental Change* 41: 142–152.

Liao, X. and Ming, J. (2019). Pressures imposed by energy production on compliance with China's 'three red lines' water policy in water-scarce provinces. *Water Policy*, 21, pp. 38–48.

Liao, X., Wei, Y. and Ma, X. (under preparation). *Water use for desulfurization in China's coal power sector.*

Ministry of Water Resources. (2013). *Regulations on implementing water resources evaluation for the planning and development of large coal power industrial clusters*. Beijing, China: Ministry of Water Resources.

National Development and Reform Commission (NDRC). (2004). *Requirements on the planning and construction of coal power plants* (in Chinese). Available at: www.nea.gov.cn/2012-01/04/c_131262602.htm.

National Development and Reform Commission (NDRC). (2015). *Intended nationally determined contribution*. Beijing, China: NDRC.

National People's Congress. (1991). *The eighth five-year plan*. Beijing, China: National People's Congress.

National People's Congress. (2001). *The tenth five-year plan*. Beijing, China: National People's Congress.

National People's Congress. (2006). *The eleventh five-year plan*. Beijing, China: National People's Congress.

National People's Congress. (2011). *The twelfth five-year plan*. Beijing, China: National People's Congress.

Organisation for Economic Co-Operation and Development (OECD). (2010). *Pricing water resources and water and sanitation services: OECD studies on water*. Paris, France: OECD Publishing.

Qin, Y., Curmi, E., Kopec, G. M., Allwood, J. M. and Richards, K. S. (2015). China's energy-water nexus assessment of the energy sector's compliance with the "3 red lines" industrial water policy. *Energy Policy*, 82, pp. 131–143.

Sanders, K. T. (2015). Critical review: Uncharted waters? The future of the electricity-water nexus. *Environmental Science & Technology*, 40, pp. 51–66.

Shi, W. X., Liu, C., Chen, W., Hong, J. L., Chang, J. C., Dong, Y. and Zhang, Y. L. (2017). Environmental effect of current desulfurization technology on fly dust emission in China. *Renewable and Sustainable Energy Reviews*, 72, pp. 1–9.

State Council of China. (2007). *Notice on closing down small thermal electric power generating units* (in Chinese). Beijing, China. Available at: www.sdpc.gov.cn/zcfb/zcfbqt/200701/t20070131_115037.html.

State Economic and Trade Commission of China. (2001). *Guide for water saving of thermal power plant (DL/T 783–2001)* (in Chinese). Beijing, China: State Economic and Trade Commission of China.

Webster, M., Donohoo, P. and Palmintier, B. (2013). WaterCO2 trade-offs in electricity generation planning. *Nature Climate Change*, 3, pp. 1029–1032.

World Bank. (under preparation). *Water knowledge note: Benchmarking China's water performance to global standards*.

World Economic Forum. (2017). *China just switched on the world's largest floating solar power plant*. Geneva, Switzerland: World Economic Forum.

World Resources Institute. (2015). *A closer look at China's new climate plan (INDC)*. Available at: www.wri.org/blog/2015/07/closer-look-chinas-new-climate-plan-indc.

Wu, L. and Huo, H. (2014). Energy efficiency achievements in China's industrial and transport sectors: How do they rate? *Energy Policy*, 73, pp. 38–46.

Zhang, C., Zhong, L. J., Fu, X. T. and Zhao, Z. N. (2016). Managing scarce water resources in China's coal power industry. *Environmental Management*, 57(6), pp. 1188–1203. doi:10.1007/s00267-016-0678-2.

9 Opportunities for and obstacles to institutional change

Overview of Chinese political system and water-energy administrations

China's political system is a multilayered hierarchical system. China's government stems from the National People's Congress (NPC) downward. The NPC is the highest organ of state power in the People's Republic of China (PRC). The State Council, headed by the Chinese Premier, is the chief executive organ of the Chinese government, which is responsible for drafting policies, regulations and laws to be reviewed and adopted by the NPC, as well as implementing those policies. After the institutional reform in 2018, the State Council compromises 27 ministries. The ones that are concerned with water management in the power sector mainly include the National Development and Reform Commission, Ministry of Water Resources, Ministry of Ecology and Environment and Ministry of Natural Resources.

China's political system is characterized by fragmented authoritarianism with two lines of structuring governmental agencies: (1) Territorial (often referred to as horizontal) – the territorial governments are set up from the central government to provincial, municipal governments and counties, towns and villages. They are responsible for all issues within their respective administrative area. (2) Functional or sectoral (often referred to as vertical) – ministries or ministry-level agencies are set up under the State Council to be concerned with particular sectors (e.g. water (Ministry of Water Resources), ecology and environment (Ministry of Ecology and Environment)). These ministries have corresponding bureaus at each territorial level (e.g. provincial water bureaus) (Ma and Ortolano, 2000).

The political rank determines power relationships in the Chinese political system. Every governmental agency is given a rank (Tsang and Kolk, 2010). Under the State Council, there are commissions, ministries, administrations and offices that are of ministry-rank (buji). It should be noted that ministry-level commissions (e.g. National Development and Reform Commission (NDRC)), usually hold higher authority than the other ministries since they have the mandates to formulate policies across geographical or economic sectors. There are another two ranks below the ministry-rank, namely, department-level (tingji or siji) and division-level (chuji). Figure 9.1 takes the water sector as an example to illustrate the two lines of governmental agencies in China and their respective rank.

This ranking system is critical in determining the geometry of policy bargaining. Negotiation takes place between institutions and individuals of equal rank. Agencies within the same rank cannot issue binding orders to each other (Andrews-Speed, 2010). A provincial government has the same rank as a ministry. Hence, a minister cannot hold authority over or issue binding orders to a provincial governor. Some large state-owned enterprises (SOE), including many large energy SOEs, also hold a ministerial or departmental level rank (Cunningham, 2007).

Sector agencies report to their superior unit along the territorial line. Chinese provinces are financially highly autonomous (Ding, McQuoid and Karayalcin, 2019). Provincial governments decide upon appointment of leaders and funding of provincial sector departments (Figure 9.2). Although sector ministries are tasked to formulate sector policies and regulations, policy implementation and enforcement fall on the shoulder of provincial sector departments. Although provincial sector departments are supposed to report to both their superior agencies (i.e. provincial governments and sector ministries), due to the funding and human resources arrangements, provincial governments have much larger power over provincial sector departments than the sector ministries, which create room for local protectionism (Lieberthal, 1997).

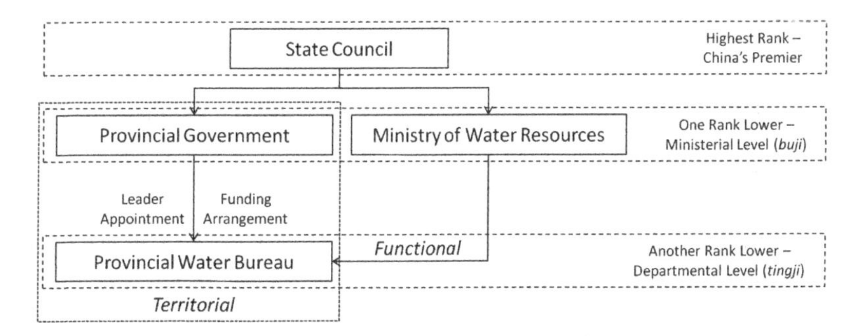

Figure 9.1 Exemplary structure of China's administrative system

Note: Taking the water sector as an example, arrows lead from order-giving party to order-receiving party.

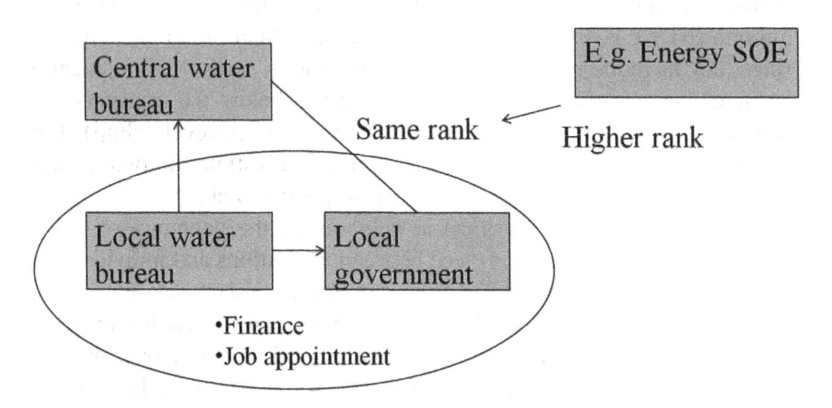

Figure 9.2 Reporting lines of local water bureaus

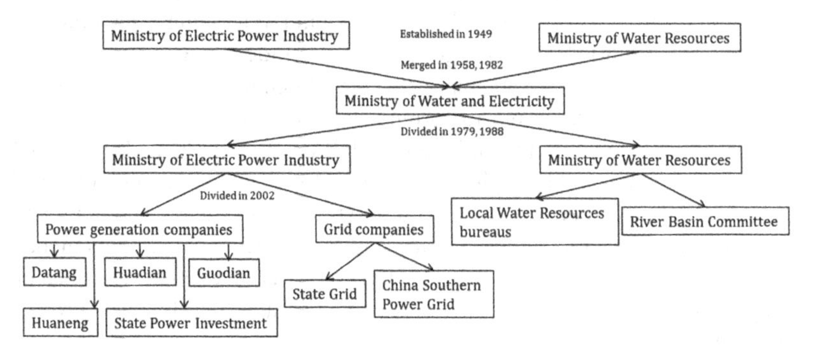

Figure 9.3 Institutional evolution of water management in the coal power sector in China

China's water and energy governance systems have gone through substantial changes. As shown in Figure 9.3, the Ministry of Water Resources and Ministry of Electric Power Industry were established when the People's Republic of China was founded in 1949. In 1958, they were merged into the Ministry of Water and Electricity, which was divided again in 1979 and combined again in 1982 and finally divided in 1988 until now. Such historical organizational changes demonstrated that the interaction between the two sectors has long been acknowledged by Chinese policymakers, while managing them in an integrated manner involves tremendous complexity and difficulties. Later, in 2002, the Bureau of Electricity, which used to solely govern electricity-related issues, was divided into five state-owned electricity production corporations, namely, Huadian, Datang, Huaneng, Guodian (Guodian and another company Shenhua merged into China Energy in 2017) and State Power Investment, and two state-owned grid companies, State Grid and China Southern Power Grid, which are responsible for electricity dispatch, transmission, distribution and supply (Li, 2003).

Water management in the power sector involves multiple agencies. Water resources are managed by the Ministry of Water Resources in China. The Ministry of Water Resources develops water utilization plans and provincial water quotas that are executed by local water resource bureaus. The NDRC and local DRCs are responsible for short-, medium- and long-term power construction and development plans, while the Ministry of Industry and Information Technology and local Commissions of Economy and Information Technology are responsible for electricity production coordination, demand management and so forth.

The National Energy Commission (NEC) is the highest-level interagency commission headed by the Chinese Premier in charge of national energy strategies. Ministers of a number of relevant ministries, including NDRC, MWR, MEE and Ministry of Natural Resources (MNR), are members of the Commission. A vice ministry-level agency, National Energy Administration, is set up within NDRC and entrusted by the NEC to take charge of the general coordination of energy development strategies and policy decisions and to ensure energy security. The Department of Electric Power under NEA specifically undertakes tasks in thermoelectric power sector planning and so forth.

The Ministry of Water Resources is tasked to manage and protect water resources. According to Chinese laws (i.e. the Constitution and the 2002 Water Law), water resources are national properties, and the state government has the responsibility to administer and protect them. The MWR has established six river basin authorities as its dispatch agencies to manage and protect water resources. At present, there are six river basin commissions and one lake commission as shown in Figure 9.4. River Basin Authorities are also responsible for tasks including inter-jurisdiction coordination, basin water allocation and so forth. Water quotas set by river basin authorities are usually used as a reference, and their main responsibility is to coordinate conflicts over water uses between different provinces within the same river basin.

Large SOEs dominate the coal power-producing industry. Large SOEs and government-controlled firms normally have some degree of natural monopoly or market power, which were developed as part of the planned economy (Caldecott et al., 2017). The state-owned Assets Supervision and Administration Commission performs investor responsibilities and manages the state-owned assets of the enterprises under the supervision of the

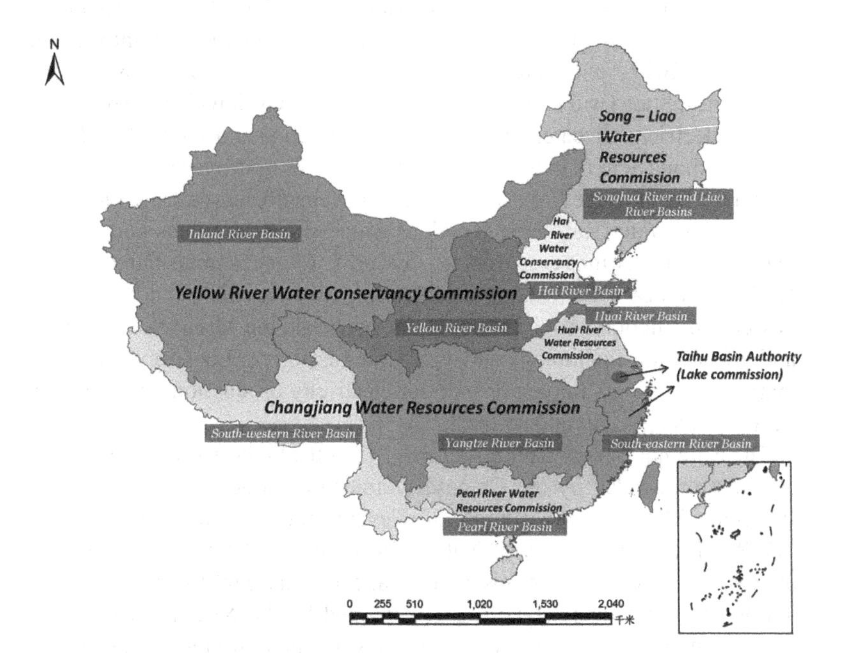

Figure 9.4 River Basin Commission Jurisdictions

central government and enhances the management of the state-owned assets. The five largest SOEs (i.e. Huaneng, Datang, Huadian, China Energy and State Power Investment Corporation) in the electric power sector together account for more than 50% of the national capacity. All of the five SOEs hold vice ministry-level rank. Those large SOEs retain considerable influence at the highest levels of government and play an important role in policymaking in their respective sectors (Andrews-Speed, 2004).

Construction of new coal power plants requires permits from a number of different governmental agencies. As shown in Figure 9.5, before a coal power plant can be constructed, its owner (i.e. power production companies that are often SOEs) is required to obtain relevant permits from a number of different governmental agencies, including the economic planning agency (i.e. Development and Reform Commission), land management agency, environmental protection agency, water resources management agency and so forth. Depending on their generating capacity, while large power plants are required to obtain permits from the ministries, small and medium power plants are allowed to apply for permits from respective provincial departments (Greenpeace, 2014). Water withdrawal permits are issued by

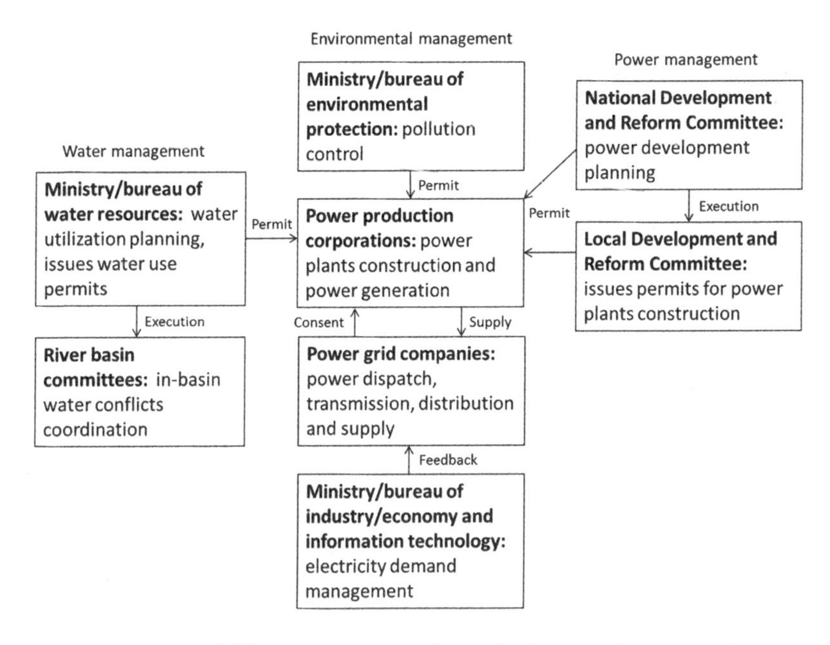

Figure 9.5 Roles of different governmental agencies in approving new coal power plants

the Ministry of Water Resources based on the water resource assessment included in the feasibility report that ensures the power plants are able to comply with all the relevant water-related regulations. However, it has been argued that the regulating power of the water withdrawal permit has not been strongly enforced (News Cableabc, 2013).

Obstacles for paradigm shift in improving water management in the coal power sector

Local protectionism impedes enforcement of water sector policies. Energy development and construction of new coal power plants often generate local economic and social benefits, including offering a large number of employment opportunities. For instance, Guangdong's electric power sector employed around 0.3 million people in 2014 (Caldecott et al., 2017). When SOEs reach agreements with local governments on constructing new power plants, the local water departments have very limited power to enforce sector regulations. There have been cases where power plants have been moving into construction before obtaining the water withdrawal permit.

It has also created barriers for the further development of renewable energies. China has heavily subsidized its renewable energy sector, which has set up numerous wind farms and solar PV farms. However, its electricity grid and transmission systems have lagged behind. There are currently many wind power and solar power assets in China's northern regions that have been abandoned due to network congestion. The problem is no longer technical but institutional. Lack of inter-provincial coordination or benefit-sharing mechanisms have resulted in provinces that do not have abundant renewable resources, preferring to develop and use their own coal power plants instead of receiving electricity generated in wind or solar-abundant regions (Qi et al., 2019).

The mismatched boundaries of energy planning and water management have potentials to result in tragedy of the commons. China's current water allocations are administered by provinces instead of by river basins. Only a few water-stressed river basins, such as the Yellow River Basin, have formulated water allocation plans (World Bank, 2019). Since energy planning and coal power construction are managed by provincial and local governments, lack of inter-provincial coordination within the same basin has resulted in the classic tragedy of the commons (i.e. unsustainable development and over-exploitation of the shared water resources).

Last but not least, economic instruments are insufficient to reflect the importance of water resources. As mentioned in the previous chapter, pricing water properly is able to incentivize water-saving practices. However, not only are water prices low in China compared to other countries, but many provinces also apply non-volumetric tariff rates to coal power plants' water use, which means power plants' water uses are charged according to the unit of electricity generated (RMB/KWh). This arrangement does not provide the coal power sector with incentives to adopt advanced water-saving technologies and improve their water use efficiencies (Zhang et al., 2016), which is aggravated by the fact that local water departments sometimes are not powerful enough to collect those fees (Fang et al., 2002).

Opportunities for improving water management in the coal power sector

China is embarking upon a transition to a more balanced and sustainable economic growth model with an emphasis on sustainable resource management. China's national 12th Five-Year Plan (2011–2015) highlighted the need for 'green development' and committed to establish a resource-saving and environmentally friendly society (National People's Congress 2011). China continued these commitments in key policy documents, including the 13th Five-Year Plan (FYP) (2016–2020) (National People's

Congress 2016) and the 19th Party Congress Report (October 2017) (Congress of the Communist Party of China, 2019), which called for a 'beautiful China' by pursuing productivity and innovation-driven development.

A series of ministerial-level institutional reforms in 2018 reflect the government's commitment toward more holistic management of natural resources, including water. This reform process focused on taming the cross-sectoral and inter-jurisdictional challenges of managing water. It clarified responsibilities by reassigning functions of the Ministry of Water Resources to two ministries: Ministry of Ecological Environment (MEE) and the newly established Ministry of Natural Resources (MNR). Reforms of vertical environmental management have already taken place, which is aimed at tackling the problems created by local protectionism (General Office of the CPC Central Committee and General Office of the State Council, 2016). After this reform, local environmental monitoring, inspection and law enforcement are supposed to directly report to the provincial environmental department instead of local governments, which aims to enhance environmental law enforcement and forms the first step to break local protectionisms.

Since 2016, China has initiated the national river chief system aimed at improving inter-sector and inter-jurisdictional coordination in managing its water resources. The system of river and lake chiefs has created a network of individuals at local, county, and provincial levels responsible for each section of every major waterway (General Office of the State Council, 2017). This system creates a platform for collaboration under the lead of the government heads that has proven useful in coordinating trans-jurisdiction and cross-sector issues. According to a report on the enforcement of the water pollution, prevention and control law submitted in August 2019 to the Standing Committee of the NPC, China now has over 1.2 million river and lake chiefs with more than 200,000 in the Yangtze River Basin alone (NPC, 2019).

The Ministry of Water Resources started expanding inter-provincial river basin water allocation in 2011. In August 2014, a technical review of the first batch of 25 river basin water allocation plans was completed. In March 2015, MWR consulted provincial and municipal governments on the river basin water allocation programs in the Nen, Huai, Han, and 17 other rivers. In October 2016, with the approval of the State Council, the MWR officially approved the first batch of water allocation plans for inter-provincial river basins (Han, Jialing, Min, Tuo, and Chishui Rivers) (World Bank, 2019). Because provinces are located across different river basins, provincial total water withdrawal caps do not necessarily cap water withdrawals from each single basin. The basin water allocation plan is therefore expected to reduce the flexibility of local governments and to improve the regulating power of water departments in economic sectors.

In 2016, China launched a comprehensive water resource fee-to-tax reform, with Hebei, where water scarcity is daunting and groundwater over-exploitation is severe, as the first pilot province. the Ministry of Finance (MOF), the State Taxation Administration (STA) and the Ministry of Water Resources (MWR) have jointly issued the 'Interim Measures for the Pilot Reform of Water Resource Tax' for Hebei province to replace the current water resource fee system (MOF, SAT and MWR, 2016). The tax is levied on the use of surface and groundwater, with higher rates on enterprises that consume large volumes of water. Water resource fees and tax reform, backed up with new legal powers, has increased collection rates. After the first tax trial was launched for 18 months, non-agricultural water consumption in Hebei province dropped by 0.18 km^3 compared to 2016 (STA, 2017).

All of these aforementioned institutional reforms in China's water sector offer significant potentials to improve the coordinated water-energy management in China. A paradigm shift must be facilitated to change the tradition governance system where water and energy are managed in silo without considering their impacts on the other. Water sustainability needs to be incorporated into the country's energy development plans and policies. Potential water constraints and risks should also be considered by energy investors and practitioners.

References

Andrews-Speed, P. (2004). *Energy policy and regulation in the people's Republic of China*. London: Kluwer Law.

Andrews-Speed, P. (2010). *The institutions of energy governance of China*. Paris, France: The Institut Francais des relations internationales.

Caldecott, B., Tulloch, D. J., Bouveret, G., Liao, X., Dericks, G. and Kruitwagen, L. (2017). *Political economy implications of stranded coal-fired assets in China: Smith school of enterprise and the environment*. Oxford: University of Oxford Press.

Congress of the Communist Party of China. (2019). *19th party congress report* (in Chinese). Beijing, China: Congress of the Communist Party of China.

Cunningham, E. A. (2007). *China's energy governance: Perception and reality*. Cambridge: MIT Center for International Studies.

Ding, Y., McQuoid, A. and Karayalcin, C. (2019). Fiscal decentralization, fiscal reform, and economic growth in China. *China Economic Review*, 53, pp. 152–167.

Fang, G., Xu, L., Wu, W. and Tan, W. (2002). A survey of water resources fee levying and management in China (in Chinese). *Water Resources and Economics of Journal*, 20(5), pp. 33–37.

General Office of the CPC Central Committee and General Office of the State Council. (2016). *Guiding opinions on environmental monitoring, inspection and law enforcement vertical management reform below provincial levels* (in Chinese).

Beijing, China: General Office of the CPC Central Committee and General Office of the State Council.
General Office of the State Council. (2017). *Opinions on fully promoting the river chief system*. Beijing, China: General Office of the State Council.
Greenpeace. (2014). *New coal power plants application process and stakeholder analysis* (in Chinese). Beijing, China: Greenpeace.
Li, G. (2003). *Separation of power plants from the grids – a theme of the reform in China power industry*. Paris, France: OECD Publishing, NEA.
Lieberthal, K. (1997). China's governing system and its impact on environmental policy implementation. *China Environment Series*, 1, pp. 3–8.
Ma, X. and Ortolano, L. (2000). *Environmental regulation in China: Institutions, enforcement, and compliance*. Lanham, MD: Rowman & Littlefield.
Ministry of Finance, State Taxation Administration, Ministry of Water Resources. (2016). *Interim measures for the pilot reform of water resource tax*. Beijing, China: Ministry of Finance, State Taxation Administration, Ministry of Water Resources.
National People's Congress. (2011). *The twelfth five-year plan*. Beijing, China: National People's Congress.
National People's Congress. (2016). *The thirteenth five-year plan*. Beijing, China: National People's Congress.
National People's Congress. (2019). *Report of the law enforcement inspection group of the standing committee of the National People's Congress on examining the implementation of the water pollution prevention and control law of the people's Republic of China*. Beijing, China: National People's Congress.
News Cableabc. (2013). *The case of Tienan Liu exposes the value chain of thermoelectric power projects application* (in Chinese). Available at: http://news.cableabc.com/domestic/20130823000580.html.
Qi, Y., Dong, W., Dong, C. and Huang, C. (2019). Understanding institutional barriers for wind curtailment in China. *Renewable and Sustainable Energy Reviews*, 105, pp. 476–486.
State Taxation Administration. (2017). *Hebei: Water resources tax exceed 1.8 billion Yuan in one year* (in Chinese). Available at: www.chinatax.gov.cn/n810219/n810739/c2725731/content.html.
Tsang, S. and Kolk, A. (2010.) The evolution of Chinese policies and governance structure on environment, energy and climate. *Environmental Policy and Governance*, 20, pp. 180–196.
World Bank. (2019). *Watershed: A new era of water governance in China*. Washington, DC: World Bank.
Zhang, C., Zhong, L., Fu, X. and Zhao, Z. (2016). Managing scarce water resources in Chinas coal power industry. *Environmental Management*, 57(6), pp. 1188–1203.

10 Conclusions

China has the largest population and the second-largest economy in the world, which requires the world's largest amount of electricity generation. About 75% of China's electricity is generated from coal-fired power plants, whose operation has given rise to a number of different environmental challenges that are of regional to global concerns. Burning coal has emitted substantial amounts of pollutants to the air that have resulted in severe health consequences. Furthermore, China has the highest total carbon emissions of any country in the world, also largely owing to its coal power production. The development of China's coal power sector is therefore of global concern.

In the early 2000s, several instances of coal power plants being shut down or curtailed due to water-related issues, including water shortages and high water temperatures around the world (e.g. France, California), raised water concerns in the energy community. Water withdrawal for coal power production accounts for more than 50% of the national total water withdrawal in the US. The coal power sector is the second-largest water user in China and highly depends on reliable water supplies. As illustrated in Chapter 2, water is required at various processes at coal power plants, from coal washing, steam generating and desulfurization to cooling purposes, among which the majority is used as a cooling medium to dissipate residual heat from steam exiting the turbines. Hence, coal power plants' water use intensity is predominantly determined by the cooling technology used. Open-loop cooling systems withdraw the largest amount of water, but over 90% is returned back to the natural water bodies. Closed-loop cooling systems withdraw substantially less water but consume almost all of it through evaporation during its cooling cycles. Air-cooling systems reduce water use compared to the other two technologies but at the expense of reduced energy efficiency and increased carbon emissions. Furthermore, as we described in detail in Chapter 3, that water is not only used by power plants' on-site activities but also at various different stages of the whole upstream fuel cycle. Large amounts of water are used for coal mining and washing, as well as for cultivating coal mine props in the agricultural sector.

Global studies show that the usable thermal power capacity is projected to decrease under the changing climate due to potentially changed water regimes, including water availability and variability (Van Vliet et al., 2016). China is particularly vulnerable to such risks due to its large number of coal power plants built on inland waterways. First of all, China's water resources per capita amount to only one-fourth of the global average. Furthermore, its water endowments are mismatched with its coal reserves and social-economic development. The majority of China's coal reserves are located in its arid and semi-arid northern regions, where most coal power plants choose to locate to save coal transportation costs, whereas water is abundant in the southern regions. Closed-loop cooling technology prevails throughout the whole country, while open-loop cooling systems are popular in the south next to large rivers or on the coast. Air-cooling systems have proliferated in the northern regions, particularly after the Chinese government issued a policy in 2004 requiring all new coal power plants in water-scarce areas to deploy air-cooling systems.

In Chapter 4 we examined the spatiotemporal patterns of water use, including both water withdrawal and water consumption, in China's coal power plants as well as their historical trajectories and potential future changes. Nationally, total water withdrawal and consumption by coal power plants in China have increased from 40.75 and 1.25 billion m^3, respectively, in 2000 to 124.06 and 4.86 billion m^3 in 2015. Water withdrawal is higher in water-abundant regions, including the east and south, where open-loop cooling systems are most prevalent, whereas water consumption is higher in the north where closed-loop cooling systems are widely deployed. Per capita electricity production has been identified as the primary driver of the historical water use increases at China's coal power plants. However, China's per capita electricity consumption is still below the world average and far below that of some developed countries. Scenario analysis reveals that future water withdrawals by coal power plants are still expected to grow unless energy transition to renewable sources or cooling technology reconfigurations are facilitated. Improving energy efficiency and transforming the energy structure offer potentials to halve the water use by coal power plants.

China's coal power plants face water shortage risks under both the current situation and future climate change. In Chapter 5, water availability was assessed on a monthly basis by a physically based hydrological model in the current period and in the 2050s under different carbon emission scenarios. Currently 20% of China's coal power plants face low-flow water shortage risks, defined as potentially insufficient low-flow river runoffs to meet coal power plants' water withdrawal demands, from November to June, with 10% facing water shortage risks during the rest of the year. Such risks are particularly pronounced in the north grid, where over 60% of its capacity (more than 100 GW) is facing water shortage risks in April. Sadoff et al. (2015) pointed out that north China stands out as having the highest water shortage risks for power production in the world. Fortunately, wetter conditions that are projected under future climate change are expected to lower such risks except in China's northwestern Inland Rivers Basin where reduced water availability is projected. However, such estimation is made under the assumption that coal power plants will be running at the current utilization rate. If the existing coal power plants are running at higher utilization rates, additional plants will face water shortage risks throughout the country, particularly in the east and south grids where water withdrawal requirements are high. Electricity demands management, energy transition away from coal and increasingly adopting air-cooling systems. These are all feasible options to reduce such risks.

Understanding the water impacts of China's power sector requires consideration of inter-provincial electricity transmission (Chapter 6). In order to utilize the rich coal resources in the western inland regions to fuel development on the eastern coast, China has constructed extensive transmission infrastructure systems. In 2014, 6.31 and 0.622 billion m^3 of water was withdrawn and consumed, respectively, to produce all the inter-provincial transmitted electricity in China. However, on the other hand, 26.41 billion m^3 of water withdrawal and 0.64 billion m^3 of water consumption was avoided in electricity-importing provinces by importing electricity. On a national level, because coal power plants in the electricity-exporting regions use more water-efficient technologies, inter-provincial electricity transmissions have resulted in 20.10 billion m^3 of water withdrawal savings and 0.021 billion m^3 of water consumption savings. Furthermore, China has started to construct another 12 long-distance transmission lines, among which nine are designed to transmitt electricity entirely generated from coal, two are designed to transmit a mix of electricity generated from wind power and coal power facilities and the other one is designed to transmit hydropower from Yunnan province in the southwest. Similarly, these 11 lines are expected to realize 58.21 to 164.52 million m^3 of water consumption and 17.90 to 18.16 billion m^3 of water withdrawal savings on the national level, depending on the different mixes of wind power and coal power utilizations. It should be noted that inter-provincial electricity transmissions have resulted in two consequences that merit close examination and special attention in the future: (1) Inter-provincial electricity transmissions impose increasing pressures on water resources in the electricity-exporting regions due to the significant amounts of water used for coal power productions. For instance, although Shanxi province is a water-scarce province with only 330 m^3 of water resources per person, it is still exporting large amounts of 'virtual' water resources embodied in the electricity it exports to other provinces; (2) inter-provincial electricity transmission exposes electricity demands in electricity-importing regions to external water shortage risks. Particularly, there are three megalopolises rapidly growing in China – Bohai Sea Economic Rim (where China's political center Beijing is located), Yangtze River Delta (where China's financial center Shanghai is located) and the Pearl River Bay area (where the coordinated development of Guangdong, Hong Kong and Macau has been defined as a national strategy); these three regions are concentrated with large inward migrations and are home to rapid economic growth. These three regions are also recipients of large amounts of inter-provincial electricity transmissions whose electricity demands are therefore exposed to water shortage risks in the inland areas further in the west.

In Chapter 7 we examined the water constraints for two alternative energy sources of coal (i.e. hydropower and concentrated solar power (CSP)). Both hydropower and CSP are regarded as low-carbon alternatives to replace coal power plants for climate change mitigation reasons. Endowed with the second-largest hydropower potential in the world, China has developed the world's largest hydropower capacity, which provides about 17% of its annual electricity consumption. Climate change studies have predicted that although China's national gross hydropower potential is expected to increase under climate change, mainly due to the projected river runoff increases in the current dry northern regions, China's existing hydropower facilities, mostly located in the southern and central regions, are expected to face water constraints due to projected river discharge reductions. Such trends are in line with global projections. Concentrated solar power (CSP) is an upcoming technology that is increasingly being adopted and offers potentials to utilize China's abundant solar energy in the northwestern regions to produce low-carbon electricity. However, as CSP facilities require water uses that are on par with coal power plants and China's solar power is primarily endowed in water-deficient regions, the future development of CSP is also expected to be constrained by available water resources. Although dry cooling for CSP is able to mitigate such conflicts, the energy efficiency penalty resulting from lower cooling efficiency renders CSP facilities using dry cooling economically unfeasible. Further studies are required to research technological innovations for developing CSP facilities under water constraints.

Water use by the power sector in China raises challenges of policy coordination (Chapter 8). Policies that are intended to improve the water performance of coal power plants include water withdrawal standards, groundwater restrictions, requirements to use air-cooling systems and unconventional water sources. Moreover, policies in the coal power sector that are intended to improve the sector's environmental performance may generate either co-benefits of or a trade-off with water conservation. Improving energy efficiency delivers co-benefits of water savings while desulfurization leads to increased water use. Particularly, China has committed to global climate change mitigation and has been acting strongly on its energy transition away from coal. Developing renewable energies such as wind power and solar PV offers substantial potentials to also cut the power sector's water impacts. On the other hand, water sector policies also regulate or even impose constraints on the further development of coal power plants. In 2011, China issued the 'Three Red Lines' policy, setting targets on total water use, water efficiency improvement and water quality improvement at the provincial level for 2020 and 2030. Many large coal-producing provinces in northern China are facing challenges in meeting their targets for both provincial total water use and industrial water efficiency improvement. The coal power industry in those provinces is expected to employ water-saving technologies and promote water-saving practices.

Water management in the coal power sector is posing emergent challenges for the development of both China's energy and water sectors, which call for further policy reforms. However, as pointed out in Chapter 9, there exist several institutional obstacles to more efficient and sustainable management of water resources for the power sector. China's government system is hierarchically set up along both territorial and functional lines. However, because provinces are financially autonomous and manage personnel and funding arrangements of provincial and local sector departments, the implementation of sector policies is largely weakened by local protectionism. Traditionally, provincial governments prioritize economic growth and social benefits over environmental sustainability. Environmental and natural resources departments, including water departments, at both provincial and local levels usually have very limited political power in regulating behaviors that may cause environmental problems, which is even hardened by the high political rank and power held by large energy SOEs. However, China is embarking upon a transition toward a more balanced and sustainable growth model, reflecting the increasing demand for good environmental quality and natural resources sustainability with the growth of people's income. Such transition opens a number of different opportunities to promote institutional changes to better manage water resources in the coal power sector. To name a few, the river chief system designates top governmental officials as river chiefs and is expected to enhance cross-sector and cross-jurisdictional coordination in managing water resources; the vertical institutional reform requires local environmental inspection, monitoring and law enforcement to directly report to the provincial environmental department instead of local government, which is aimed at breaking the bottleneck of local protectionism; water resources fee-to-tax reform increases the water resource tax collection rate with legal authority in order to effectively incentivize water-saving technologies and practices.

China's demand for electricity is expected to continue to grow in the future to a point where per capita consumption comes into line with high-income countries. It is clear that continuing to fuel electricity production with coal will not be consistent with the country's commitments to curb carbon emissions. China already has the largest hydropower production in the world and has invested heavily in solar and wind power generation. A transition away from coal-fired power production is already taking place, but in the meantime, the majority of the country's electric power is generated from coal-fired power plants. Besides their emissions of carbon dioxide and other harmful air pollutants, these plants are responsible for the second-largest (after agriculture) quantity of freshwater withdrawals in China, so they have a significant impact on the aquatic environment. This book has sought to examine and quantify the impacts of China's power sector on water resources and the risks to that sector from water scarcity. We have seen how over-exploitation of water resources by the power sector in the past has been a consequence of misaligned incentives and inadequate regulations and enforcement, which are now being reformed thanks to more rigorous water policies. Deeper scientific understanding, better regulation and a transition to cleaner sources of electric power should contribute to more sustainable management of China's water resources in future.

References

Sadoff, C. W., Hall, J. W., Grey, D., Aerts, J. C. J. H., Ait-Kadi, M., Brown, C., Cox, A., Dadson, S., Garrick, D., Kelman, J., McCornick, P., Ringler, C., Rosegrant, M., Whittington, D. and Wiberg, D. (2015). Securing Water, Sustaining Growth: Report of the GWP/OECD Task Force on Water Security and Sustainable Growth, University of Oxford, Oxford.

Van Vliet, M. T. H., Wiberg, D., Leduc, S. & Riahi, K. (2016). Power-generation system vulnerability and adaptation to changes in climate and water resources. *Nature Climate Change*, 6, pp.375–380.

Index

Note: Page numbers in *italic* indicate a figure and page numbers in **bold** indicate a table on the corresponding page.

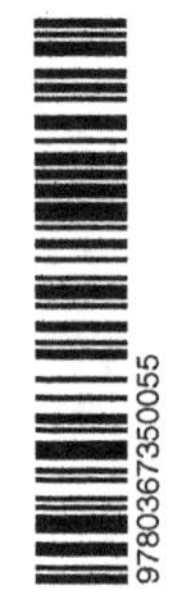